Synthesis Lectures on Data Management

Editor
M. Tamer Özsu, ***University of Waterloo***

Synthesis Lectures on Data Management is edited by Tamer Özsu of the University of Waterloo. The series will publish 50- to 125 page publications on topics pertaining to data management. The scope will largely follow the purview of premier information and computer science conferences, such as ACM SIGMOD, VLDB, ICDE, PODS, ICDT, and ACM KDD. Potential topics include, but not are limited to: query languages, database system architectures, transaction management, data warehousing, XML and databases, data stream systems, wide scale data distribution, multimedia data management, data mining, and related subjects.

Data Stream Management
Lukasz Golab and M. Tamer Özsu
2010

Access Control in Data Management Systems
Elena Ferrari
2010

An Introduction to Duplicate Detection
Felix Naumann and Melanie Herschel
2010

Privacy-Preserving Data Publishing: An Overview
Raymond Chi-Wing Wong and Ada Wai-Chee Fu
2010

Keyword Search in Databases
Jeffrey Xu Yu, Lu Qin, and Lijun Chang
2009

Data Stream Management

Lukasz Golab and M. Tamer Özsu

www.morganclaypool.com

ISBN: 9781608452729 paperback
ISBN: 9781608452736 ebook

DOI 10.2200/S00284ED1V01Y201006DTM005

A Publication in the Morgan & Claypool Publishers series
SYNTHESIS LECTURES ON DATA MANAGEMENT

Lecture #5
Series Editor: M. Tamer Özsu, *University of Waterloo*
Series ISSN
Synthesis Lectures on Data Management
Print 2153-5418 Electronic 2153-5426

Data Stream Management

Lukasz Golab
AT&T Labs—Research, USA

M. Tamer Özsu
University of Waterloo, Canada

SYNTHESIS LECTURES ON DATA MANAGEMENT #5

ABSTRACT

In this lecture many applications process high volumes of streaming data, among them Internet traffic analysis, financial tickers, and transaction log mining. In general, a data stream is an unbounded data set that is produced incrementally over time, rather than being available in full before its processing begins. In this lecture, we give an overview of recent research in stream processing, ranging from answering simple queries on high-speed streams to loading real-time data feeds into a streaming warehouse for off-line analysis.

We will discuss two types of systems for end-to-end stream processing: Data Stream Management Systems (DSMSs) and Streaming Data Warehouses (SDWs). A traditional database management system typically processes a stream of ad-hoc queries over relatively static data. In contrast, a DSMS evaluates static (long-running) queries on streaming data, making a single pass over the data and using limited working memory. In the first part of this lecture, we will discuss research problems in DSMSs, such as continuous query languages, non-blocking query operators that continually react to new data, and continuous query optimization. The second part covers SDWs, which combine the real-time response of a DSMS by loading new data as soon as they arrive with a data warehouse's ability to manage Terabytes of historical data on secondary storage.

KEYWORDS

Data stream Management Systems, Stream Processing, Continuous Queries, Streaming Data Warehouses

Data Stream Management

Contents

CHAPTER 1

Introduction

Many applications process high volumes of streaming data. Examples include Internet traffic analysis, sensor networks, Web server and error log mining, financial tickers and on-line trading, real-time mining of telephone call records or credit card transactions, tracking the GPS coordinates of moving objects, and analyzing the results of scientific experiments. In general, a data stream is a data set that is produced incrementally over time, rather than being available in full before its processing begins. Of course, completely static data are not practical, and even traditional databases may be updated over time. However, new problems arise when processing unbounded streams in nearly real time. In this lecture, we survey these problems and their solutions.

1.1 OVERVIEW OF DATA STREAM MANAGEMENT

We use network monitoring as a running example. Figure 1.1 illustrates a simple IP network with high-speed routers and links in the core, and hosts (clients and servers) at the edge. A large network contains thousands of routers and links, and its core links may carry many thousands of packets per second; in fact, optical links in the Internet backbone can reach speeds of over 100 million packets per second [Johnson et al., 2008]. The traffic flowing through the network is itself a high-speed data stream, with each data packet containing fields such as a timestamp, the source and destination IP addresses, and ports. Other network monitoring data streams include real-time system and alert logs produced by routers, routing and configuration updates, and periodic performance measurements. Examples of performance measurements are the average router CPU utilization over the last five minutes and the number of inbound and outbound packets of various types over the last five minutes. Understanding these data streams is crucial for managing and troubleshooting a large network. However, it is not feasible to perform complex operations on high-speed streams or to keep transmitting Terabytes of raw data to a data management system. Instead, we need scalable and flexible end-to-end data stream management solutions, ranging from real-time low-latency alerting and monitoring, ad-hoc analysis and early data reduction on raw streaming data, to long-term analysis of processed data.

We will discuss two complementary techniques for end-to-end stream processing: Data Stream Management Systems (DSMSs) and Streaming Data Warehouses (SDWs). Figure 1.2 compares DSMSs, SDWs and traditional data management systems in terms of data rates on the y-axis, and query complexity and available storage on the x-axis. Traditional data warehouses occupy the bottom left corner of the chart, as they are used for complex off-line analysis of large amounts of relatively static, historical data (warehouse updates are usually batched together and applied during

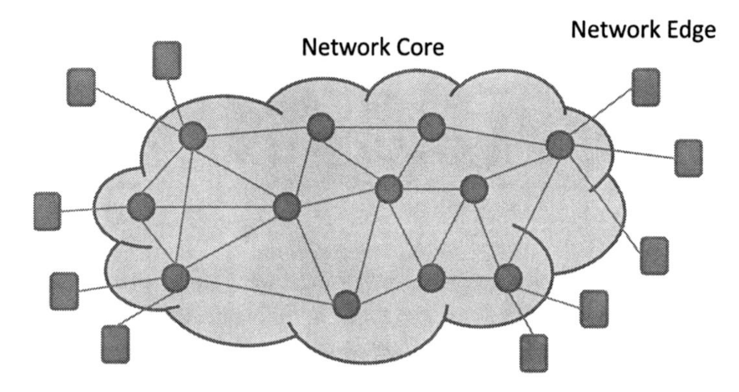

Figure 1.1: A simple network.

downtimes). Database Management Systems (DBMSs) handle somewhat more dynamic workloads, consisting of ad-hoc queries and data manipulation statements, i.e., insertions, updates and deletions of a single row or groups of rows. On the other hand, DSMSs lie in the top right corner as they evaluate *continuous queries* on data streams that accumulate over time. In applications such as troubleshooting a live network, the data rates may be so high that only the simplest continuous queries that require very little working memory and per-tuple processing are feasible, such as simple filters and simple aggregates over non-overlapping windows. Furthermore, we may have to return approximate answers over an incrementally maintained sample of the stream. Other streaming sources, such as sensor networks, may have lower data rates. If so, it may be possible to answer more complex queries on-line, such as aggregates over *sliding windows* of recent data. Even then, a DSMS needs to do all the data processing in main memory due to the high overhead of disk I/O. Finally, SDWs [Golab et al., 2009a], also known as *Active Data Warehouses* [Polyzotis et al., 2008], combine the real-time response of a DSMS (by attempting to load and propagate new data across materialized views as soon as they arrive) with a data warehouse's ability to manage Terabytes of historical data on secondary storage. In network monitoring, an SDW may store traffic streams that have been pre-aggregated or otherwise pre-processed by a DSMS, as well as various network performance and configuration feeds that arrive with a wide range of inter-arrival times, e.g., once a minute to once a day.

Table 1.1 summarizes the differences between DBMSs and DSMSs. The processing model of a DBMS is pull-based or query-driven, in which short-lived queries are executed once (when posed) by "pulling" the current state of the database. In a DSMS, the processing model is push-based or data-driven, where continuous queries are issued once and generate new results as new data arrive. That is, rather than evaluating transient (asked once and then "forgotten") queries on persistent data, a DSMS evaluates persistent queries on transient, append-only data. For example, in network monitoring, some continuous queries may run for several minutes or several hours to troubleshoot specific problems; other queries gather routine performance measurements and may

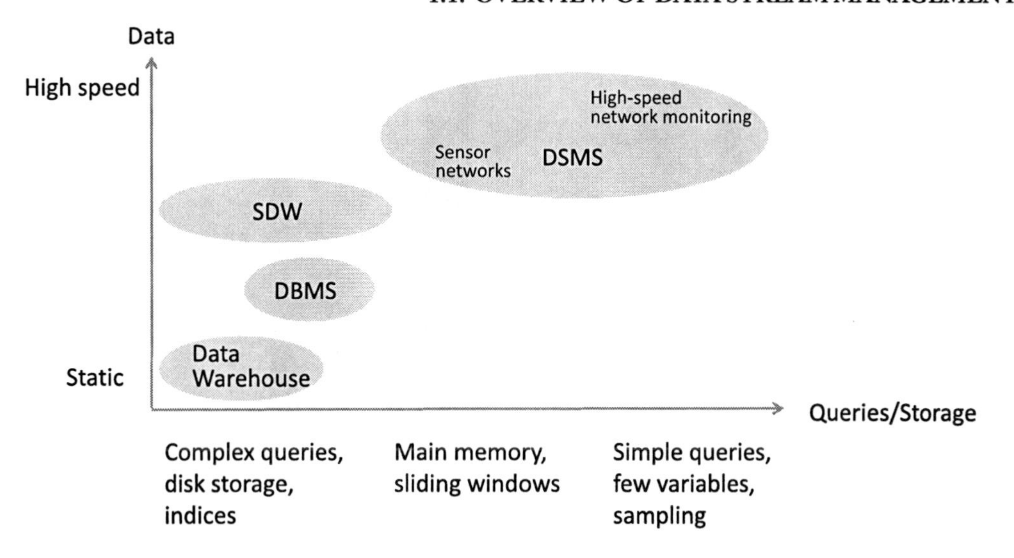

Figure 1.2: Comparison of Data Stream Management Systems and Streaming Data Warehouses with traditional database and warehouse systems.

Table 1.1: Summary of differences between a DBMS and a DSMS.

	DBMS	DSMS
Data	persistent relations	streams, time windows
Data access	random	sequential, one-pass
Updates	arbitrary	append-only
Update rates	relatively low	high, bursty
Processing model	query-driven (pull-based)	data-driven (push-based)
Queries	one-time	continuous
Query plans	fixed	adaptive
Query optimization	one query	multi-query
Query answers	exact	exact or approximate
Latency	relatively high	low

run for weeks or even months. Query processing strategies are also different. A DBMS typically has (random) access to all the data during query execution and employs a fixed query plan to generate exact answers. In contrast, a DSMS must process a data stream in one sequential pass, using limited working memory, and may adjust its query plans in response to changes in the data. Since DSMSs must support many continuous queries that run for a long period of time, multi-query optimization becomes important. Furthermore, DSMSs may generate approximate answers and must do so with strict latency requirements.

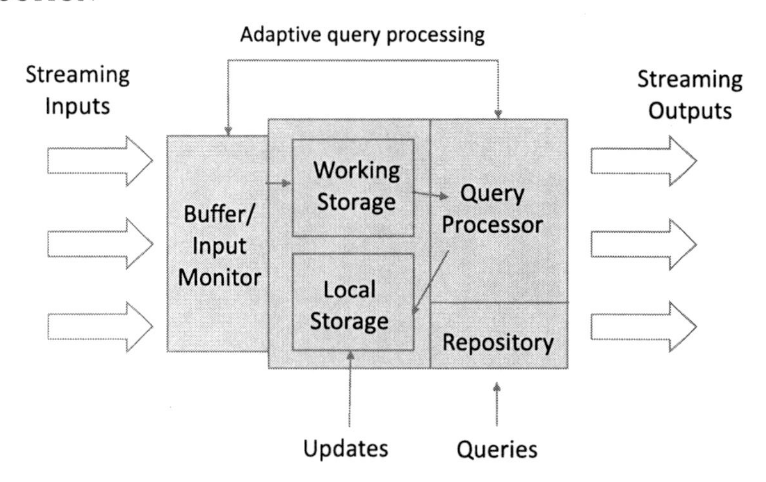

Figure 1.3: Abstract reference architecture of a DSMS.

Figure 1.3 shows an abstract system architecture of a DSMS. On the left, an input buffer captures the streaming inputs. Optionally, an input monitor may collect various statistics such as inter-arrival times or drop some incoming data in a controlled fashion (e.g., via random sampling) if the system cannot keep up. The working storage component temporarily stores recent portions of the stream and/or various summary data structures needed by queries. Depending on the arrival rates, this ranges from a small number of counters in fast RAM to memory-resident sliding windows. Local storage may be used for metadata such as foreign key mappings, e.g., translation from numeric device IDs that accompany router performance data to more user-friendly router names. Users may directly update the metadata in the local storage, but the working storage is used only for query processing. Continuous queries are registered in the query repository and converted into execution plans; similar queries may be grouped for shared processing. While superficially similar to relational query plans, continuous query plans also require buffers, inter-operator queues and scheduling algorithms to handle continuously streaming data. Conceptually, each operator consumes a data stream and returns a modified stream for consumption by the next operator in the pipeline. The query processor may communicate with the input monitor and may change the query plans in response to changes in the workload and the input rates. Finally, results may streamed to users, to alerting or event-processing applications, or to a SDW for permanent storage and further analysis.

Next, we summarize the differences between a traditional data warehouse and a SDW in Table 1.2. The fundamental difference is the higher frequency and asynchronous nature of updates—rather than refreshing the entire warehouse periodically, a SDW attempts to load new data as they arrive so that any applications or triggers that depend on the data can take immediate action. Furthermore, a SDW makes recent and historical data available for analysis. This requires, among other things, a fast Extract-Transform-Load (ETL) process and efficient update propagation across layers of materialized views in order to keep up with the inputs.

Table 1.2: Summary of differences between a traditional data warehouse and a SDW.

	Traditional data warehouse	SDW
Update frequency	low	high
Update propagation	synchronized	asynchronous
Data	historical	recent and historical
ETL process	complex	fast, light-weight

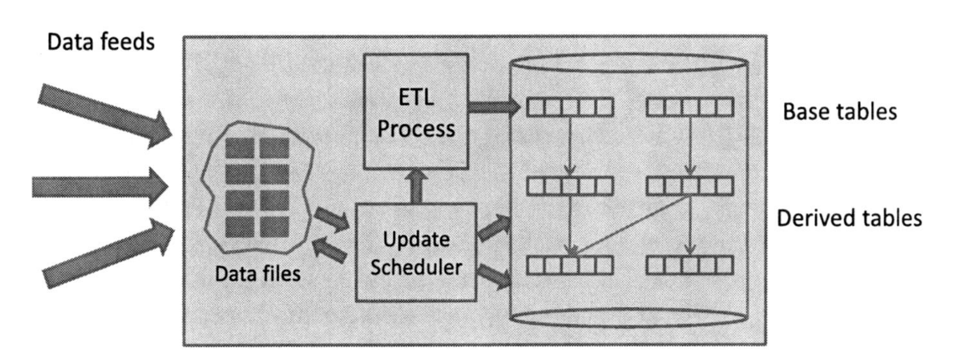

Figure 1.4: Abstract reference architecture of a SDW.

In Figure 1.4, we illustrate an abstract system architecture of a SDW. Data streams or feeds arrive periodically from various sources, often in the form of text or gzipped files. An update scheduler decides which file or batch of files to load next. The data then pass through an ETL process, as in traditional data warehouses. Examples of ETL tasks include unzipping compressed files, and simple data cleaning and standardization (e.g., converting strings to lower or upper case or converting timestamps to GMT). Base tables are sourced directly from the raw files, while derived tables correspond to materialized views (over base or other derived tables). Base and derived tables are usually partitioned by time so that arrivals of new data only affect the most recent partitions. Furthermore, users may specify (or the system may automatically infer) *partition dependencies* between derived tables and their sources. For instance, if a base table B is partitioned into 5-minute chunks and a derived table D is partitioned by hour (e.g., perhaps D computes hourly aggregates on B), then we know that the most recent partition of D depends on the twelve most recent partitions of B. Partition dependencies enable efficient update propagation in an SDW, as we can identify specific derived table partitions that need to be updated in response to an update of a base table. In addition to choosing raw files to load, the scheduler is also responsible for choosing which derived table to update next.

1.2 ORGANIZATION

The remainder of this lecture is organized as follows. Chapter 2 covers DSMSs. We begin with a discussion of data models and query languages that take into account the temporal and unbounded nature of data streams. We will discuss concepts such as temporal windows, semantics of continuous queries that are issued once and updated over time, and SQL-based as well as workflow-based approaches to querying data streams. We then discuss DSMS query processing techniques for the data-driven computation model, including designing operators that continually react to new data and scheduling these operators to minimize various metrics such as throughput or memory usage. We also discuss novel query optimization issues in DSMSs, including adapting to changes in system conditions during the lifetime of a continuous query, load shedding and multi-query optimization.

We remark that instead of designing a new system for data stream management, we can construct one using a DBMS along with triggers and application code to simulate data-driven processing and continuous queries [Arasu et al., 2004b; Babcock et al., 2002; Stonebraker et al., 2005]. One problem with this approach is the overhead associated with DBMS functionalities that are not essential for stream processing, such as recovery logging (data streams are not stored persistently) or concurrency control (data streams are usually append-only and are generated by a single "writer"). Another issue is that triggers do not scale to streaming speeds and rich continuous queries [Abadi et al., 2003].

Next, in Chapter 3, we discuss SDWs. A fundamental difference between an SDW and a traditional warehouse is that the former must load new data and propagate changes across materialized views as soon as possible. We will overview recent approaches to speeding up the Extract-Transform-Load (ETL) process so that new data can be loaded efficiently and new update scheduling and propagation techniques for maximizing data freshness. We will also discuss query processing issues in an SDW.

Finally, we conclude in Chapter 4.

We note of existing related books and surveys on on-line data stream algorithms [Muthukrishnan, 2005], data stream mining [Aggarwal, 2007; Garofalakis et al., 2010], data stream algorithms, languages and systems [Chaudhry et al., 2005], and query processing on data streams [Babcock et al., 2002; Chakravarthy and Jiang, 2009; Golab and Özsu, 2003a]. To the best of our knowledge, this is the first survey of end-to-end data stream management, covering real-time stream processing in DSMSs and historical data stream collection in SDWs.

We conclude this section with a list of selected DSMSs and SDWs that we will discuss in the remainder of this lecture. We will focus on systems that extend the relational model to continuous queries on streaming data. However, we remark that there exists a class of related systems that perform event processing and pattern matching on streams [Agrawal et al., 2008; Demers et al., 2007; Wu et al., 2006]. These systems are based on the finite state machine model, which is outside the scope of this lecture.

- **Aurora** [Abadi et al., 2003; Balakrishnan et al., 2004; Carney et al., 2003; Tatbul et al., 2003] and its distributed version **Borealis** [Abadi et al., 2005; Balazinska et al., 2005; Hwang et al.,

2007; Tatbul et al., 2007; Xing et al., 2005, 2006] are based on the workflow model. Rather than specifying continuous queries in an SQL-like language, users build boxes-and-arrows dataflow diagrams that are then translated into continuous query plans.

- **CAPE** [Liu et al., 2005; Rundensteiner et al., 2004; Wang and Rundensteiner, 2009; Zhu et al., 2004] is a general-purpose DSMS that emphasizes adaptive query processing and quality-of-service guarantees on the streaming results.

- **GS Tool** [Cormode et al., 2004; Cranor et al., 2003; Golab et al., 2008b; Johnson et al., 2005a,b,c, 2008] is a light-weight DSMS designed for high-speed network monitoring. It supports a limited subset of SQL, as well as user-defined aggregate functions, sampling, and distributed operation to handle high data rates.

- **NiagaraCQ** [Chen et al., 2000, 2002] is an early systems for evaluating a large number of continuous queries consisting of selections, projections and joins. Its main focus is on scalability and multi-query optimization.

- **Nile** [Ghanem et al., 2007, 2010; Hammad et al., 2004] is a general-purpose DSMS, focusing on issues such as sliding window query processing, multi-query optimization, and maintaining materialized views of the results of continuous queries.

- **PIPES** [Krämer, 2007; Krämer and Seeger, 2004, 2009] is a data stream processing framework implemented in Java, providing streaming operators based on a temporal algebra, and a runtime system components such as operator scheduling and query re-optimization.

- **STREAM** [Arasu et al., 2006; Babcock et al., 2002, 2004a,b; Babu and Widom, 2004; Motwani et al., 2003] is a general-purpose stream engine that employs a novel SQL-based continuous query language.

- **Stream Mill** [Bai et al., 2006] is a flexible DSMS, whose unique feature is that it allows user-defined functions to be written in SQL rather than an external language.

- **TelegraphCQ** [Chandrasekaran and Franklin, 2004; Chandrasekaran et al., 2003; Deshpande and Hellerstein, 2004; Reiss and Hellerstein, 2005] is a stream system based on PostgreSQL DBMS and focuses on adaptive query processing and combining real-time and historical data.

- **DataDepot** [Golab et al., 2009a,b] is a streaming data warehouse developed at AT&T Labs-Research. It features real-time data loading, update scheduling, and efficient update propagation via partition dependencies.

CHAPTER 2

Data Stream Management Systems

In this chapter, we discuss DSMSs, including stream data models, query languages and semantics, and query processing and optimization.

2.1 PRELIMINARIES

2.1.1 STREAM MODELS

For the purpose of this lecture, data streams have the following properties:

- They are sequences of records, ordered by arrival time or by another ordered attribute such as generation time (which is likely to be correlated with, but not equivalent to, the arrival time), that arrive for processing over time instead of being available a priori;
- They are produced by a variety of external sources, meaning that a DSMS has no control over the arrival order or the data rate;
- They are produced continually and, therefore, have unbounded, or at least unknown, length. Thus, a DSMS may not know if or when the stream "ends".

We distinguish between *base streams* produced by the sources and *derived streams* produced by continuous queries and their operators [Arasu et al., 2006]. In either case, we model individual stream items as relational tuples with a fixed schema. For instance, an Internet traffic stream (more specifically, a TCP or UDP packet stream) may have the following schema, where `size` indicates the number of bytes of data contained in each packet and `timestamp` is the packet generation time:

```
<timestamp, source IP address, source port, destination IP address,
destination port, size>.
```

Since there may be multiple connections between the same pair of nodes (IP addresses and ports) over time, the first five fields form a key. We may also tag each tuple with its arrival time in addition to the generation time; this is useful for measuring the response time of the DSMS (i.e., the time it takes to produce results in response to the arrival of a new tuple) [Abadi et al., 2003]. On the other hand, a stream of router CPU utilization measurements may have the following schema, with new measurements arriving from each router every five minutes (here, timestamp and router ID are a key):

`<timestamp, router ID, cpu_usage>`.

Some models require a stream to be totally ordered, whereas others do not distinguish among items having the same value of the ordering attribute. Recently, a unified model was proposed that allows both alternatives [Jain et al., 2008].

In practice, base streams are almost always append-only in the sense that previously arrived items (i.e., those with previously seen keys) are never modified; in the remainder of this lecture, we assume that every base stream is append-only. However, derived streams may or may not be append-only. For instance, consider the following query, $Q1$, over an Internet traffic stream S (assume that `timestamp` is the packet generation time measured in seconds, similar to a Unix timestamp):

```
Q1:  SELECT    minute, source_IP_address, SUM(size) AS total_traffic
     FROM      S
     GROUP BY  timestamp/60 AS minute, source_IP_address
```

At the end of each one-minute window, this query computes the total traffic originating from each source IP address during the given window (we will discuss continuous query semantics and languages in more detail later in this chapter). Thus, `minute` and `source_IP_address` are a key for the output stream of this query. However, since packets may arrive late, incorrect `total_traffic` values may be returned if they are computed *eagerly* right at the end of every minute, say, according to the local clock. One solution is to produce revised answers [Ryvkina et al., 2006] that take late arrivals into account, meaning that the output stream may contain updates to previously seen `total_traffic` values. Note that this example is meant only to illustrate that derived streams may not be append-only. As we will discuss later in this chapter, there are other ways of dealing with out-of-order arrivals, such as buffering.

Data streams often describe underlying "signals". Let $S : [1...N] \rightarrow \mathbb{R}$ be a signal, represented as a mapping from a discrete domain to the reals. For example, one signal carried by an Internet packet stream maps each source IP address to the total number of bytes sent from that address (i.e., $N = 2^{32}$ since an IP address is 32 bits long); another maps each destination IP address to the total number of bytes received by it. There are at least four ways in which a data stream can represent a signal [Gilbert et al., 2001; Hoffmann et al., 2007; Muthukrishnan, 2005]:

- In the *aggregate model*, each stream item contains a range value for a particular value in the domain of S.
- In the *cash register model*, each stream item contains a partial non-negative range value for a particular value in the domain of S. Hence, to reconstruct the signal A, we need to aggregate the partial range values for each domain value.
- The *turnstile model* generalizes the cash register model by allowing the partial range values to be negative.
- In the *reset model*, each stream item contains a range value for a particular value in the domain of S, which replaces all previous range values for this domain value.

The Internet packet stream examples correspond to the cash register model since the underlying signals (e.g., total traffic per source and total traffic per destination) are computed by aggregating data from individual packets, while router CPU measurements are examples of the reset model since new measurements effectively replace old ones (assuming that the domain of the signal consists of all the routers being monitored). Alternatively, in the router CPU example, a signal whose domain elements are labeled with the router id and the 5-minute measurement window corresponds to the aggregate model: each domain value occurs exactly once in the stream, assuming that there is exactly one measurement from each router in each 5-minute window. The turnstile model usually appears in derived streams, e.g., a difference of two cash-register base streams.

2.1.2 STREAM WINDOWS

From the system's point of view, it is infeasible to store an entire stream. From the user's point of view, recently arrived data may be more useful. This motivates the use of windows to restrict the scope of continuous queries. Windows may be classified according the following criteria.

1. *Direction of movement:* Fixed starting and ending points define a *fixed window*. Moving starting and ending points (either forward or backward) creates a *sliding window*. The starting and ending points typically slide in the same direction, although it is possible for both to slide outwards, starting from a common position (creating an expanding window), or for both to slide inwards, starting from fixed, positions (creating a contracting window). One fixed point and one moving point define a *landmark window*. Usually, the starting point is fixed, and the ending point moves forward in time. There are a total of nine possibilities as the beginning and ending points may independently be fixed, moving forward or moving backward.

2. *Definition of contents:* Logical or *time-based* windows are defined in terms of a time interval, e.g., a time-based sliding window may maintain the last ten minutes of data. Physical (also known as *count-based* or *tuple-based*) windows are defined in terms of the number of tuples, e.g., a count-based sliding window may store the last 1000 tuples. When using count-based windows in queries with a `GROUP BY` condition, it may be useful to maintain separate windows of equal size for each group, say, 1000 tuples each, rather than a single window of 1000 tuples; these are referred to as *partitioned windows* [Arasu et al., 2006]. The most general type is a *predicate window* [Ghanem et al., 2006], in which an arbitrary logical predicate (or SQL query) specifies the contents. Predicate windows are analogous to materialized views. For example, consider an on-line auction that produces three types of tuples for each item being sold: a "begin" tuple, zero or more "bid" tuples, followed by an "end" tuple that is generated when the item has been sold, presumably to the highest bidder. Assume that each tuple contains a timestamp, an item id, a type (being, bid or end), and other information such as the bid amount. A possible predicate window over this auction stream keeps track of all items that have not yet been sold. New items enter the window when their "begin" tuples appear; sold items "fall out" of the window when their "end" tuples appear.

3. *Frequency of movement:* By default, a time-based window is updated at every time tick, and a count-based window is updated when a new tuple arrives. A *jumping window* is updated every k ticks or after every kth arrival. Note that a count-based window may be updated periodically, and a time-based window may be updated after some number of new tuples have arrived; these are referred to as *mixed jumping windows* [Ma et al., 2005]. If k is equal to the window size, then the result is a series of non-overlapping *tumbling windows* [Abadi et al., 2003].

In practice, tumbling windows, such as the one-minute windows in query $Q1$ from Section 2.1.1, are popular due to the simplicity of their implementation—at the end of each window, the query resets its state and starts over. Forward-sliding windows (time-based and count-based) are also appealing due to their intuitive semantics, especially with joins and aggregation, as we will discuss below. However, sliding windows are more difficult to implement than tumbling windows; over time, a continuous query must insert new tuples into a window and remove *expired* tuples that have fallen out of the window range.

2.2 CONTINUOUS QUERY SEMANTICS AND OPERATORS

2.2.1 SEMANTICS AND ALGEBRAS

In a DSMS, queries are issued once and run continuously, incrementally producing new results over time. Let Q be a continuous query, whose inputs are one or more append-only data streams and zero or more relations. For now, assume that relations are static throughout the lifetime of the query (we will relax this assumption in Section 2.2.4). Let $Q(\tau)$ be the result of Q at time τ. Intuitively, we want Q to produce "correct" answers at any point in time, taking into account all the data that have arrived so far.

Definition 2.1 At any time τ, $Q(\tau)$ must be equal to the output of a corresponding traditional database query, in which all references to streams are replaced by relations constructed from prefixes of the streams up to time τ.

Optionally, the output may be refreshed periodically, in which case the variable τ in the above definition ranges over the refresh times rather than each time tick.

Two types of continuous query algebras have been proposed in the literature, both based on relational algebra. In a *stream-to-stream* algebra, each operator consumes one or more streams (and zero or more relations) and incrementally produces an output stream [Cranor et al., 2003; Krämer and Seeger, 2009]. In a *mixed* algebra [Arasu et al., 2006; Ghanem et al., 2010], there are three sets of operators: those which produce a relation from a stream (e.g., sliding windows), those which produce a relation from one or more input relations (i.e., the standard relational algebraic operators), and those which produce a stream from a relation. Conceptually, at every time tick, an operator converts its input to relations, computes any new results, and converts the results back a stream that can be consumed by the next operator. Since the converted relations change over time, a natural way of switching back to a stream is to report the difference between the current result and the result computed one time tick ago. This is similar to computing a set of changes (insertions and/or deletions) required to update a materialized view.

2.2.2 OPERATORS

We now introduce common continuous query operators, deferring a discussion of continuous query processing and optimization in Section 2.4. First, we need to define two important concepts: monotonicity and non-blocking execution.

Definition 2.2 A continuous query or continuous query operator Q is monotonic if $Q(\tau) \subseteq Q(\tau')$ for all $\tau \leq \tau'$.

For example, simple selection over a single stream or a join of several streams are monotonic (recall our assumption that base streams are append-only). To see this, note that when a new tuple arrives, it either satisfies the (selection or join) predicate or it does not, and the satisfaction condition does not change over time. Thus, at any point in time, all the previously returned results remain in $Q(\tau)$. On the other hand, continuous queries with negation or set difference are non-monotonic, even on append-only streams.

Definition 2.3 A continuous query or continuous query operator Q is non-blocking if it does not need to wait until it has seen the entire input before producing results.

Some operators have blocking and non-blocking implementations; of course, only the latter are used in DSMSs since we do not know when we will have seen the entire input. For instance, traditional SQL queries with aggregation are blocking since they scan the whole relation and then return the answer. However, on-line aggregation [Hellerstein et al., 1997; Law et al., 2004], where partial answers are incrementally returned as they are computed over the data seen so far, is non-blocking. Note that Definitions 2.2 and 2.3 are related: Q is monotonic if and only if it is non-blocking [Law et al., 2004].

The simplest continuous query operators are *stateless*; examples include duplicate-preserving projection, selection, and union. These operators process new tuples on-the-fly without storing any temporary results, either by discarding unwanted attributes (projection) or dropping tuples that do not satisfy the selection condition (technically, the union operator temporarily buffers the inputs to ensure that its output stream is ordered). Figure 2.1(a) shows a simple example of selection (of all the "a" tuples) over the character stream $S1$.

A non-blocking, pipelined join (of two character streams, $S1$ and $S2$) [Wilschut and Apers, 1991] is illustrated in Figure 2.1(b). A hash-based implementation maintains hash tables on both inputs. When a new tuple arrives on one of the inputs, it is inserted into its hash table and probed against the other stream's hash table to generate results involving the new tuple, if any. Joins of more than two streams and joins of streams with a static relation are straightforward extensions. In the former, for each arrival on one input, the states of all the other inputs are probed in some order

[Golab and Özsu, 2003b; Viglas et al., 2003]. In the latter, new arrivals on the stream trigger the probing of the relation [Balazinska et al., 2007].

Since maintaining hash tables on unbounded streams is not practical, most DSMSs only support window joins. Query $Q2$ below is an example of a tumbling window join (on the attribute *attr*) of two streams, $S1$ and $S2$, where the result tuples must satisfy the join predicate and belong to the same one-minute tumbling window. Similar to $Q1$, tumbling windows are created by grouping on the timestamp attribute. At the end of each window, the join operator can clear its hash tables and start producing results for the next window.

```
Q2:  SELECT    *
     FROM      S1, S2
     WHERE     S1.attr = S2.attr
     GROUP BY  S1.timestamp/60 AS minute
```

One disadvantage of $Q2$ is that matching tuples, whose timestamps are only a few seconds apart but happen to fall into different tumbling windows, will not be reported. Another option is a sliding window join [Golab and Özsu, 2003b; Kang et al., 2003], where two tuples join if they satisfy the join predicate and if their timestamps are at most one window length, call it w, apart. A sliding window join may be expressed in a similar way to $Q3$ below:

```
Q3:  SELECT    *
     FROM      S1, S2
     WHERE     S1.attr = S2.attr
     GROUP BY  |S1.timestamp - S2.timestamp| <= w
```

Alternatively, if the query language allows explicit specification of sliding windows (e.g., using the SQL:1999 RANGE keyword), $Q3$ may be expressed as follows:

```
Q4:  SELECT  *
     FROM    S1 [RANGE w], S2 [RANGE w]
     WHERE   S1.attr = S2.attr
```

Note that tumbling window joins are simpler to evaluate since we can clear the hash tables at the end of every window and start over. In contrast, sliding window joins need to maintain their hash tables over time, by inserting new tuples as they arrive and removing old tuples (we will discuss how to do this efficiently in Section 2.5.2).

Figure 2.1(c) shows a COUNT aggregate. When a new tuple arrives, we increment the stored count and append the new result to the output stream. If the aggregate includes a GROUP BY clause, we need to maintain partial counts for each group (e.g., in a hash table) and emit a new count for a group whenever a new tuple with this particular group value arrives. Since aggregates on a whole stream may not be of interest to users, DSMSs support tumbling and/or sliding window aggregates. For efficiency, window aggregates are typically implemented to return new results periodically rather than reacting to each new tuple. Query $Q1$ is an example of a tumbling aggregate (with grouping

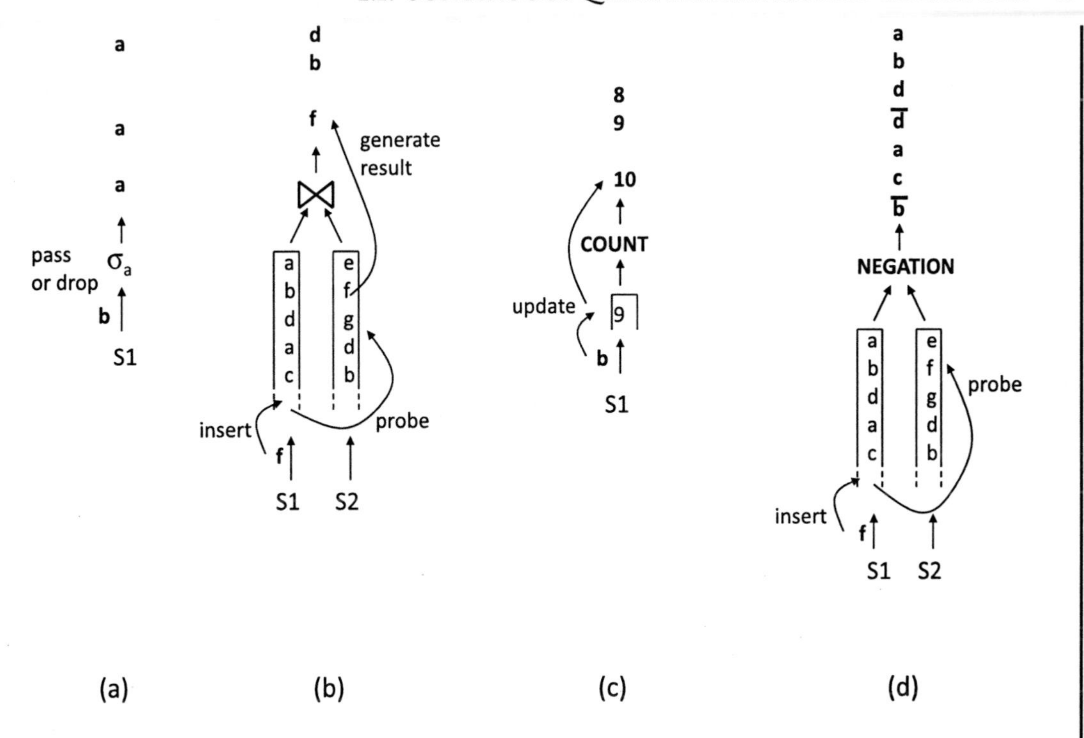

Figure 2.1: Simple continuous query operators: (a) - selection, (b) - join, (c) - count, (e) - negation.

on the source IP address) that produces a batch of results at the end of each one-minute window. We will discuss optimizing sliding window aggregates in Section 2.5.3.

Note that the type of aggregate being computed determines the amount of state that we need to store and the per-tuple processing time. An aggregate f is *distributive* if, for two disjoint multi-sets X and Y, $f(X \cup Y) = f(X) \cup f(Y)$. Distributive aggregates, such as `COUNT`, `SUM`, `MAX` and `MIN`, may be computed incrementally using constant space and time (per tuple). For instance, `SUM` is evaluated by storing the current sum and continually adding to it the values of new tuples as they arrive. Moreover, f is *algebraic* if it can be computed using the values of two or more distributive aggregates using constant space and time (e.g., `AVG` is algebraic because `AVG = SUM/COUNT`). Algebraic aggregates are also incrementally computable using constant space and time. On the other hand, f is *holistic* if, for two multi-sets X and Y, computing $f(X \cup Y)$ exactly requires space proportional to the size of $X \cup Y$. Examples of holistic aggregates include `TOP-k`, `QUANTILE`, and `COUNT DISTINCT`. For instance, we need to maintain the multiplicities of each distinct value seen so far to (exactly) identify the k most frequent item types at any point in time. This requires $\Omega(n)$ space, where n is the number of stream tuples seen so far—consider a stream with $n - 1$ unique values and one of the values occurring twice.

In practice, we can obtain approximate answers to holistic aggregates on data streams using several methods. One is to maintain a (possibly non-uniform) sample of the stream. Sampling is used for approximating `COUNT DISTINCT` [Gibbons, 2001; Pavan and Tirthapura, 2005], `QUANTILE` [Manku et al., 1999], and `TOP k` [Demaine et al., 2002; Estan and Varghese, 2002; Gibbons and Matias, 1998; Manku and Motwani, 2002] queries. Another possibility is to avoid storing frequency counters for each distinct value seen so far by periodically evicting counters having low values [Liu et al., 2006b; Manku and Motwani, 2002; Metwally et al., 2005]. This is a possible approach for computing `TOP k` queries, so long as frequently occurring values are not missed by repeatedly deleting and re-starting counters. A related space-reduction technique may also be used for approximate quantile computation where the rank of a subset of values is stored along with corresponding error bounds (rather than storing a sorted list of all the frequency counters for exact quantile calculation) [Gilbert et al., 2002; Greenwald and Khanna, 2001]. Finally, hashing is another way of reducing the number of counters that need to be maintained. Stream summaries created using hashing are referred to as *sketches*. Examples include the following.

- A Flajolet-Martin (FM) sketch [Alon et al., 1996; Flajolet and Martin, 1983] is used for `COUNT DISTINCT` queries. It uses a set of hash functions h_j that map each value v onto the integral range $[1, \ldots, \log U]$ with probability $\mathbf{Pr}[h_j(v)=l] = 2^{-l}$, where U is the upper bound on the number of possible distinct values. Given d distinct items in the stream, each hash function is expected to map $d/2$ items to bucket 1, $d/4$ items to bucket 2, and so on with all buckets above $\log d$ expected to be empty. Thus, the highest numbered non-empty bucket is an estimate for the value $\log d$. The FM sketch approximates d by averaging the estimate of $\log d$ (i.e., the largest non-zero bit) from each hash function.

- A Count-Min (CM) sketch [Cormode and Muthukrishnan, 2004] is a two-dimensional array of counters with dimensions d by w. There are d associated hash functions, h_1 through h_d, each mapping stream items onto the integral range $[1, \ldots, w]$. When a new tuple arrives with value v, the following cells in the array are incremented: $[i, h_i(v)]$ for $= 1$ to d. An estimate of the count of tuples with value v can be obtained by taking the minimum of values found in cells $[i, h_i(v)]$ for $i = 1$ to d. The approximate counts can then be used to compute `TOP k` queries.

So far, we have shown relational-like query operators and briefly discussed how a DSMS may evaluate them. Other operators in a DSMS may include the following.

- A buffered sort operator [Abadi et al., 2003] buffers the incoming stream for a specified length of time, call it l, and outputs the stream in sorted order on some attribute, typically the generation timestamp. Tuples that arrive more than l time units late are dropped.

- Many DSMSs include built-in sampling operators, either for specific techniques such as random sampling [Abadi et al., 2003; Motwani et al., 2003], or for expressing various user-defined sampling methods [Johnson et al., 2005a].

- User-defined aggregate functions (UDAFs) are supported by most DSMSs. They are often used to add application-specific functionalities to general-purpose DSMSs, such as approximate versions of complex aggregates and data mining algorithms, as described above [Cormode et al., 2004; Law et al., 2004; Muthukrishnan, 2005]. A UDAF specification requires at least two components: how to initialize any intermediate state before processing and how to update the state (and, optionally, report an up-to-date answer) when a new tuple arrives. Additionally, a UDAF over a sliding window needs to specify how to update the state when a tuple expires from the window.

2.2.3 CONTINUOUS QUERIES AS VIEWS

Up to now, we have discussed non-blocking and monotonic operators that incrementally produce new results over time. We now introduce continuous query operators whose output streams contain new results as well as updates to existing results. Conceptually, these operators produce "deltas" to a *view* of the current result (possibly, with the help of separate relation-to-stream operators), which may be materialized and maintained by an application that receives the streaming output [Ghanem et al., 2010].

We begin with (duplicate-preserving) negation, or set difference, of two character streams, $S1$ and $S2$. As illustrated in Figure 2.1(d), we want to maintain a materialized view of characters in the prefix of $S1$ that we have seen so far that do not occur in the prefix of $S2$ that we have seen so far. Initially, we return "a", "b", and "d" as soon as they arrive on $S1$ since they do not have counterparts in $S2$. When "d" arrives on $S2$, we remove its $S1$-counterpart from the result by appending a *negative tuple* [Arasu et al., 2006; Ghanem et al., 2007, 2010; Golab and Özsu, 2005] to the output, denoted by $\bar{d}$ in the illustration. When "b" arrives on $S2$ later, we need another negative tuple (denoted by $\bar{b}$) to remove its $S1$-counterpart from the result.

Another example of negative tuples in the output stream involves removing old tuples from sliding windows. Recall queries $Q3$ and $Q4$, which use windows to limit the state of the join operator. Both of these queries are monotonic and keep producing new join results over time. Now, suppose that we also want to remove old join results involving tuples that have expired from their windows (i.e., tuples more than w time units old). Figure 2.2 shows an example where the sliding windows have moved forward, causing "f" to expire from $S1$ and "c" to expire from $S2$. Given that both of these tuples have created join results in the past, the join operator must now produce corresponding negative tuples. An application can now treat the output stream as a sequence of updates to a materialized view.

As the above example shows, windows can play two roles in a DSMS: implicit (internal) and explicit (external) [Golab, 2006]. Implicit windows are used to limit the memory requirements of operators such as joins, as in $Q3$ and $Q4$. However, tuples do not expire from the result set when they expire from their windows, and, therefore, monotonic operators that use implicit windows remain monotonic. On the other hand, as illustrated in Figure 2.2, explicit windows can be thought of as non-monotonic operators that consume the input streams, and, in addition to propagating each tuple

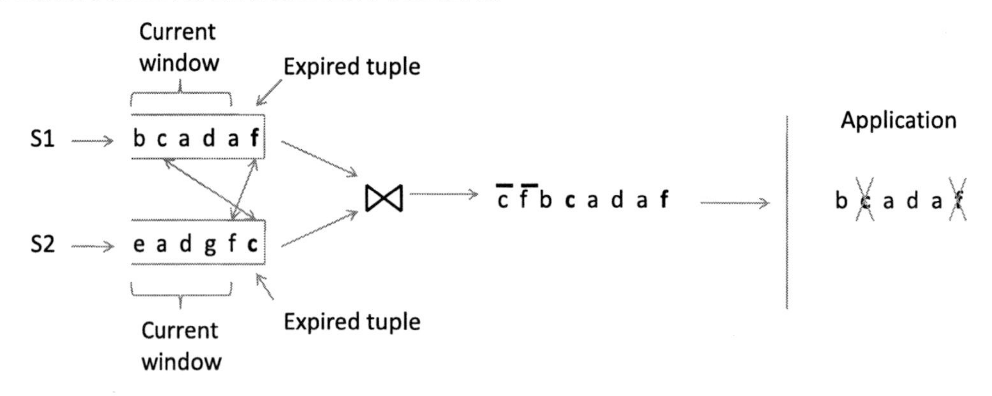

Figure 2.2: Maintaining a materialized view of a sliding window join.

to the next operator in the query plan, produce a negative tuple whenever a "positive" tuple expires. Negative tuples must then be processed by each operator in the plan to "undo" previously reported results. As a result, the output stream, when interpreted as a sequence of "deltas" to a materialized view, must correspond to the result of an equivalent traditional query executed over the current state of the sliding windows (not to the result of an equivalent query that operates over prefixes of the inputs, as in Definition 2.1). We will discuss the processing of negative tuples in Section 2.4.3.

Finally, we note that in the queries-as-views approach, the DSMS itself usually does not materialize the views. Continuous queries produce streams of updates over time, but it is up to the user or application to maintain the final result.

2.2.4 SEMANTICS OF RELATIONS IN CONTINUOUS QUERIES

The query-as-view model can also help understand the semantics of continuous queries that reference non-static relations. Suppose that we want to join a stream, bounded by a (explicit or implicit) window, with a relation. For example, the stream tuples may contain a timestamp, an internal router ID, and a performance measurement such as CPU usage; the relation may have mappings from router IDs to alphanumeric names. Clearly, we may need to insert new mappings into the relation when adding new routers to the network, and/or delete old mappings when old routers are retired (updates may be modeled as deletions followed by insertions). If we want to maintain a view of the current answer at all times, deletions from the relation must produce negative tuples so that the join operator can undo all the current results that have been produced using the deleted tuple. This is similar to producing a negative tuple in Figure 2.2, in response to expirations from the sliding windows. Similarly, after adding a tuple to the relation, we need to probe the sliding window and generate new results based on the stream tuples that have already arrived. That is, we are treating the relation as an explicit window (more specifically, an explicit predicate window—recall Section 2.1.2).

Another, more efficient, solution is to make the changes to the relation visible only to tuples that will arrive in the future, effectively treating relations as implicit windows. That is, when a new tuple is added to the relation, we do not probe the sliding window to generate join results with any stream tuples that have already arrived; instead, new stream tuples that will arrive in the future will probe the new state of the relation. Similarly, after deleting a tuple from the relation, we do not need to produce any negative tuples. These types of relation updates are referred to as *non-retroactive* [Golab and Özsu, 2005]. Though non-retroactive updates may not be appropriate in all applications, they are useful in the above network monitoring example. When an old router is removed from the relation, there is no need to expire any of its measurements that have already been generated. Similarly, when a new router is inserted into the relation *before* it starts producing measurements, then there are no prior measurements to join with this new tuple.

Note that there is an important restriction on non-retroactive updates: once we delete a row from the relation with a particular key, we cannot later add a row with the same key. For example, if we remove a router ID, but do not delete the CPU usage measurements that have been produced for this router, then inserting a new router that has the same router ID makes it seem that the old CPU usage statistics refer to the new router.

2.3 CONTINUOUS QUERY LANGUAGES

Most DSMSs employ declarative query languages with SQL-like syntax, possibly with additional keywords for windows and novel operators such as sampling or relation-to-stream conversion. One exception is SQuAl, the graphical boxes-and-arrows language used in the Aurora DSMS; however, it, too, supports non-blocking relational-like operators such as selection, projection, union, join and group-by-aggregation. Of course, as discussed earlier, although the surface syntax may be similar to SQL, the semantics and operator implementations of continuous queries are considerably different from traditional one-time queries.

We will survey selected features of SQL-like languages below. First, we remark that most DSMSs use variants of `CREATE STREAM` statements to specify the schema and source (e.g., port number) of streaming inputs. Systems that also support relational input allow tables to be defined using the standard `CREATE TABLE` command.

2.3.1 STREAMS, RELATIONS AND WINDOWS

GSQL [Cranor et al., 2003] is used by the GS Tool and is a representative of simple, purely stream-to-stream languages with monotonic operators only. It supports a restricted subset of non-blocking versions of standard relational operators and requires all the inputs (and output) to be streams. GSQL has direct support only for tumbling windows that are specified by grouping on a timestamp attribute and must be used with joins and aggregation. However, sliding window aggregates, as well as join of streams with relations, may be encoded as UDAFs. $Q1$ and $Q2$ were written in a syntax similar to GSQL.

CQL [Arasu et al., 2006], which is used by the STREAM DSMS, is a powerful language that supports streams, relations, sliding windows, and negative tuples. It contains three types of operators: relation-to-relation operators that are similar to standard relational operators, sliding windows that convert streams to time-varying relations, and three relation-to-stream operators: `Istream`, `Dstream` and `Rstream`. The `Istream` operator compares the current output of the query (represented as a relation) with the output as of the current time minus one and returns only the new results. In contrast, at any time, `Dstream` returns all the results that existed in the answer set at the current time minus one but do not exist at the current time. That is, `Dstream` returns all the negative tuples required to maintain a materialized view of the result. Finally, `Rstream` streams out the entire result at any given time.

In contrast to GSQL's tumbling-only windows, CQL supports sliding windows to convert streams to relations. Time-based windows of length N are specified with the `[RANGE N]` keyword following the reference to a stream. Count-based windows are denoted as `[ROWS N]` and partitioned windows on some attribute `attr` (recall Section 2.1.2) as `[PARTITION BY attr ROWS N]`. Windows containing only those tuples whose timestamps are equal to the current time are denoted as `[NOW]`, and a prefix of a stream up to now can be turned into a relation using `[RANGE UNBOUNDED]` or `[ROWS UNBOUNDED]`.

To illustrate CQL's windows and relation-to-stream operators, consider a simple selection over a network traffic stream S:

```
Q5:  SELECT   Istream(*)
     FROM     S  [RANGE UNBOUNDED]
     WHERE    source_IP_address = "1.2.3.4"
```

Since this query is monotonic, it suffices to output new results at every time tick (using `Istream`), calculated over all the data that have arrived so far. Since this query is stateless, another way to express is by returning all the tuples that satisfy the selection predicate at any instant of time (using `Rstream`), over a `NOW` window of tuples that have arrived at that time instant:

```
Q6:  SELECT   Rstream(*)
     FROM     S  [NOW]
     WHERE    source_IP_address = "1.2.3.4"
```

Note that using `Rstream` and unbounded windows with the above queries gives a different and arguably less desirable result—at any point in time, we repeatedly receive all the packets from IP address 1.2.3.4 that have arrived up to now. In general, the problem with repetitions in the output of `Rstream` is that other operators in the plan may not be able to process it properly, as it is neither an append-only stream of new results nor a stream of positive and negative deltas [Ghanem et al., 2010]. CQL also provides syntactic shortcuts: windows are unbounded by default; `Istream` is the default relation-to-stream operator for monotonic queries, etc. Applying these shortcuts, the above queries may be written simply as `SELECT * FROM S WHERE Source_IP_address = "1.2.3.4"`.

In CQL, joins are usually expressed using `Istream` and sliding windows. In this case, the windows are implicit since `Istream` only generates new results. To express joins over explicit windows such as those in Figure 2.2, we need to write two queries, one with `Istream` (positive deltas) and one with `Dstream` (negative deltas), and merge their results. Notably, the SyncSQL language

[Ghanem et al., 2010] used in the Nile system can return both positive and negative tuples in a single query with the SELECT STREAMED keyword.

So far, we have given examples of queries with windows whose default behaviour is to slide whenever new tuples arrive. However, jumping windows are often used with aggregation for performance reasons; additionally, users may find it easier to deal with periodic output rather than a continuous output stream. CQL as well as the ESL language [Bai et al., 2006] used in Stream Mill support aggregates over jumping windows via the SLIDE construct. For example, $Q7$ computes the total traffic originating from each source over the last five minutes, with new results returned every minute:

```
Q7:  SELECT    Rstream(source_IP_address, SUM(size))
     FROM      S  [RANGE 5 min SLIDE 1 min]
     GROUP BY  source_IP_address
```

Note that in the CQL version of this query, as shown above, we need to use Rstream to ensure that the complete result is produced every minute.

2.3.2 USER-DEFINED FUNCTIONS

Most stream query languages support user-defined functions written in an external language. ESL allows SQL-based user-defined aggregates (UDAs). An ESL UDA consists of three parts: INITIALIZE, ITERATE and TERMINATE. The first part initializes the required state, which must be stored in tables. The ITERATE statement is executed whenever a new tuple arrives. The TERMINATE statement is evaluated at the end of the input; a non-blocking UDA has an empty TERMINATE statement since it incrementally produces results as new tuples arrive.

Below, we show an example of a UDA that computes the average over a count-based tumbling window of size 200 (this example originally appeared in the ESL user manual, available at http://magna.cs.ucla.edu/stream-mill/doc/esl-manual.pdf). The current sum and count will be maintained in the state table. When the first tuple arrives, the INITIALIZE statement simply updates the state variables. The ITERATE statement is slightly more complex. At the end of every tumbling window (where Cnt % 200 = 0), it updates the state and returns the average. The first tuple of the new tumbling window (where Cnt % 200 = 1) overwrites the previous window's state.

```
AGGREGATE Tumbling_window_avg(Next Int): Real {
  TABLE State(Tsum Int, Cnt Int);
  INITIALIZE: {
    INSERT INTO State VALUES (Next, 1);
  }
  ITERATE: {
    UPDATE State
      SET Tsum = Tsum + Next, Cnt = Cnt + 1;
    INSERT INTO Return
```

```
      SELECT Tsum/Cnt FROM STATE
      WHERE Cnt % 200 = 0;
    INSERT INTO State VALUES (Next, 1)
    WHERE Cnt % 200 = 1;
  }
  TERMINATE: {}
}
```

2.3.3 SAMPLING

CQL and SQuAl provide built-in random sampling. SQuAl includes a `DROP` operator, whereas CQL provides a `SAMPLE` keyword. For example, to obtain a one percent random sample of a stream *S*, we write `SELECT * FROM S SAMPLE(1)`. GSQL provides a framework for expressing various sampling methods [Johnson et al., 2005a]. Users are required to write functions that describe the sampling technique, which are then referenced in the `HAVING`, `CLEANING WHEN` and `CLEANING BY` clauses. These clauses specify when to reset the state of the given sampling operator and which groups to return in the sample. ESL does not provide a separate sampling operator, but, similarly to GSQL, user-defined sampling algorithms may be written as UDAFs.

2.3.4 SUMMARY

Table 2.1 summarizes the continuous query languages discussed in this section in terms of their inputs and outputs, and the window types that they support; for a broader comparison, we refer the interested reader to an article by Cherniack and Zdonik [2009]. Note that GSQL natively supports only streaming inputs and only tumbling windows, but relation inputs and sliding window aggregates may be incorporated via user-defined functions.

Table 2.1: Summary of selected continuous query languages.

Language/ System	Inputs	Outputs	Supported Windows
CQL/STREAM	Streams, Relations	Positive deltas, Negative deltas, Relations	Tumbling, Sliding, Jumping
ESL/Stream Mill	Streams, Relations	Streams	Tumbling, Sliding, Jumping
GSQL/GS Tool Streams	Streams	Tumbling	
SQuAl/Aurora	Streams, Relations	Streams	Tumbling, Sliding, Jumping
SyncSQL/Nile	Streams, Relations	Streams, Deltas, Relations	Tumbling, Sliding, Jumping

2.4 STREAM QUERY PROCESSING

In this section, we discuss how a DSMS evaluates continuous queries. As in a DBMS, after a query is issued, it is translated from a declarative specification into a logical query plan containing logical operators (unless the user submits the logical query plan directly into the DSMS, as in the Aurora system). The logical plan may be statically optimized (e.g., some operators may be re-ordered) before being turned into a physical plan with concrete physical operators (e.g., a logical join operator may become a non-blocking hash join). Since multiple continuous queries may be running at any given time, we may choose to construct a single plan for all the queries and share memory and/or computation whenever possible. Furthermore, the DSMS may re-optimize the physical query plan over time.

Figure 2.3 illustrates a plan for two queries: 1) a join of streams $S1$ and $S2$ with a selection predicate on $S1$, and 2) an aggregate on $S2$. A plan for one or more continuous queries is a directed acyclic graph (DAG), with nodes corresponding to (non-blocking, pipelined) operators and edges corresponding to data flow across operators. Each plan receives data from one or more sources, and it produces one or more output streams, one per query. Note that buffers and queues are required at the inputs and between operators to handle continuously arriving data. Also, note that the $S2$-buffer is shared by two queries. When new tuples arrive on $S1$, they are retrieved from the input queue by the selection operator, and tuples that satisfy the selection predicate are passed to the next queue. The join operator must process tuples from the selection queue and the $S2$ input queue, by inserting them into the appropriate state and probing the other state. Similarly, the aggregation operator processes new tuples by updating its state and pushing an updated result on its output stream; optionally, new results may be generated periodically.

In general, an operator may produce zero, one or more result tuples for each incoming tuple. As in traditional DBMSs, we define the *selectivity* of an operator as the average number of outputs produced by a single input tuple. For example, selection predicates have selectivity between zero and one, while joins may have a selectivity greater than one. Other statistics relevant to continuous query optimization include the stream arrival rates and the per-tuple processing time of each operator. Note that knowing the stream arrival rates and the selectivity of each operator allows us to estimate each operator's output rate.

2.4.1 SCHEDULING

Suppose that the DSMS has chosen a physical plan and started to execute it. At any given time, there may be many tuples in the input and inter-operator queues, especially if the stream arrival rates are bursty. Thus, the DSMS *scheduler* must decide which tuples to process next. A simple scheduling strategy is to allocate a time slice to each operator in round-robin fashion or to execute each operator in round-robin fashion until it has processed all the tuples in its queue(s). Another simple technique, first-in-first-out, is to process one tuple at a time in order of arrival, such that each tuple is processed to completion by all the operators in the plan. However, while this strategy ensures good response time (i.e., the difference between the generation time of a result and the

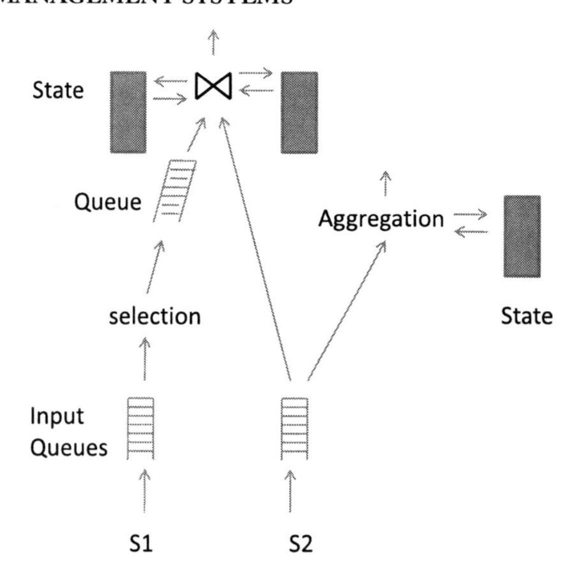

Figure 2.3: A simple continuous query plan.

arrival time of the original tuple), scheduling one tuple at a time may incur too much overhead. Instead, we can achieve low latency by scheduling operators that produce results with the highest rate (i.e., those which produce the most results per unit time) [Carney et al., 2003; Sharaf et al., 2008; Viglas and Naughton, 2002]. Similar algorithms have also appeared for minimizing average slowdown, which is the ratio between the tuple response time and the shortest possible processing time, assuming that there is no contention for resources [Sharaf et al., 2008].

On the other hand, we may want to minimize runtime memory usage by limiting the number of tuples that wait in queues. If so, it is best to give priority to operators with low processing cost and low selectivity—such operators will quickly clear out their queues and will not deposit many result tuples in the next queue [Babcock et al., 2004a].

2.4.2 HEARTBEATS AND PUNCTUATIONS

We now discuss other types of tuples that may flow through a continuous query plan, namely those which help deal with out-of-order arrivals and help "keep the pipeline moving". Since DSMSs process data arriving from external sources, some tuples may arrive late or out-of-order with respect to their generation time (a simple but often inadequate solution could be to implicitly order the stream by arrival time). Furthermore, we may not receive data from a source (e.g., a remote sensor or a router) for some time, which could mean that there are no new data to report or that the source is down. If there is a bound on the amount of time that a tuple may be late, we can simply buffer the inputs (recall Aurora's buffered sort operator from Section 2.2.2). However, the buffer time may be unknown or variable over time.

One solution to some of these problems is to embed control tuples in the data stream, called *punctuations* [Tucker et al., 2003]. A punctuation is a special tuple that contains a predicate that is guaranteed to be satisfied by the remainder of the data stream. For instance, a punctuation with the predicate `timestamp > 1262304000` guarantees that no more tuples will arrive with timestamps below the given Unix time; of course, if this punctuation is generated by the source, then it is useful only if tuples arrive in timestamp order. Punctuations that govern the timestamps of future tuples are typically referred to as *heartbeats* [Johnson et al., 2005b; Srivastava and Widom, 2004a].

Another application of punctuations is to reduce the amount of state that operators need to store and prevent operators from doing unnecessary work [Ding and Rundensteiner, 2004; Ding et al., 2004; Fernandez-Moctezuma et al., 2009; Li et al., 2006b, 2005a]. For a simple example, consider a sliding window hash join of two streams on some attribute *attr*. If a punctuation arrives on one of the input streams with the predicate `attr != a1 AND attr != a2`, we can immediately remove tuples with those *attr*-values from both hash tables. Recently, punctuations have also been used to encode dynamic access control policies on data streams [Nehme et al., 2008] and to carry user-defined annotations [Nehme et al., 2009].

In addition to punctuations and heartbeats generated by sources, the query plan itself may need to produce heartbeats to avoid pipeline stalls and delayed results [Li et al., 2008]. Consider query $Q8$, over an Internet packet stream S, which computes the total traffic to destination protocol 80 over 60-second tumbling windows.

```
Q8:  SELECT    minute, SUM(size)
     FROM      S
     WHERE     destination_port <= 80
     GROUP BY  timestamp/60 AS minute
```

A sample input is illustrated in Figure 2.4. The input attributes are `<timestamp, port, size>`; other attributes such as IP addresses are not relevant to this example. Suppose that the current tumbling window has started at time 8:01:00 and will end at time 8:01:59. From the point of view of a query operator, time advances when a new tuple appears in its queue. At time 8:01:45, the SUM operator has seen the four tuples illustrated on the right (with port numbers between 1 and 80). The next tuple arrives at 8:02:00 and signals the end of the current tumbling window; recall our assumption of ordered arrival. However, this tuple is not passed to the SUM operator, so it does not know that the window has ended until the next tuple arrives at time 8:02:15. Only then can the aggregate operator output the sum. On the other hand, consider a strategy where the selection operator sends a heartbeat up to the SUM operator if it has not propagated any results for some period of time, say, ten seconds. In this case, time advances whenever a regular tuple or a heartbeat arrives. Here, the 8:02:15 tuple triggers a heartbeat with the same timestamp, which indicates to the SUM operator that the current tumbling window has ended.

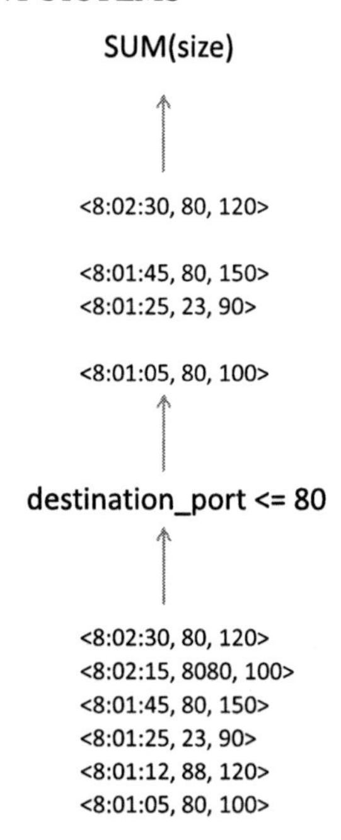

Figure 2.4: A continuous query with selection and tumbling window aggregation.

2.4.3 PROCESSING QUERIES-AS-VIEWS AND NEGATIVE TUPLES

In addition to "regular" tuples, punctuations and heartbeats, a continuous query plan may need to process negative tuples if operating in the queries-as-views mode (recall Section 2.2.3). Negative tuples may be implemented by adding a "sign" field to the schema of the stream and propagating this field through all the query operators to the output stream. As we will show, most operators require new physical variants to handle negative tuples [Ghanem et al., 2007; Golab and Özsu, 2005].

Consider the query plan shown in Figure 2.5, which evaluates a sliding window join over explicit windows. That is, at any time, the output stream must contain all the deltas that are required to construct a view of the current join result, taking into account only those tuples that are currently in the windows. The figure shows how a tuple that has expired from the Stream 1 window is processed by all the operators. New tuples are routed to the corresponding explicit window operators. These are physical operators that store the window and propagate all the incoming tuples as well as negative tuples up to the next operator. A count-based explicit window may be implemented as a circular array; each new tuple evicts the oldest tuple in the window, and the evicted tuple is passed to the next operator in the plan in the form of a negative tuple. A time-based explicit window may be implemented as a linked list, with insertions to the head and expirations from the tail. Note that time-based windows may contain a different number of tuples at different times, and that expirations

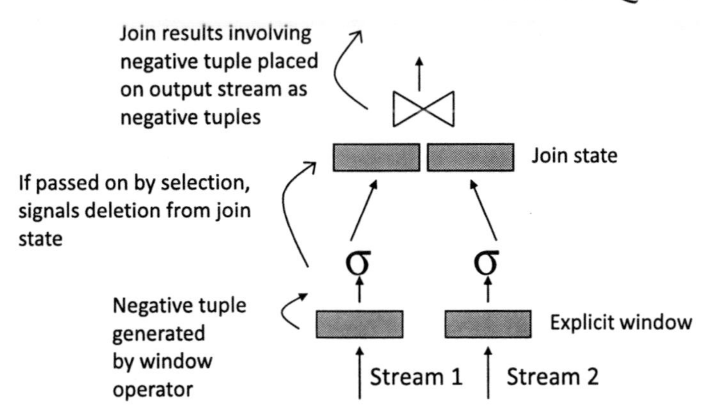

Figure 2.5: Maintaining a view over a sliding window join using negative tuples.

are independent from arrivals of new tuples. Thus, an explicit time-based window operator needs to periodically check if the oldest tuple has expired, and, if so, generate a corresponding negative tuple.

The negative tuples that have been generated by explicit window operators must then be processed by the remaining operators. Selection is an example of an operator that processes negative tuples in exactly the same way as regular tuples—it passes them to the next operator if they satisfy the selection predicate, or it drops them otherwise. The join operator also handles negative tuples in the same way as regular tuples to "undo" previous results. Note that a single negative tuple consumed by the join operator may result in zero, one or more negative tuples on the output stream, just as a single regular tuple may produce zero or many join results. Furthermore, the join operator needs to remove the corresponding regular tuple from its state (e.g., hash table) so that this tuple does not produce any new results in the future.

Now suppose that the query plan in Figure 2.5 includes an aggregation operator on top of the join. Some aggregates can easily handle negative tuples. For example, `COUNT` simply decrements the stored count for each negative tuple seen. On the other hand, while `MAX` is easy to compute on an unbounded stream or a tumbling window (we simply keep track of the largest value so far), it becomes harder when negative tuples or expirations are involved. In Figure 2.6, we give an example of finding the largest value in a sliding window, with tuples and their values shown on a time axis. At time t_1, the window spans all but the youngest tuple (with a value of 61), which has not yet arrived. The maximum at this time is 75. At time t_2, the youngest tuple arrives, but its value is smaller than the maximum, so the answer does not change. However, at time t_3, the window slides past the oldest tuple with a value of 75, i.e., the corresponding negative tuple arrives. At that time, we need to find the new maximum value, which is 73. In the worst case, this requires storing the entire window.

(A simple optimization to reduce the memory requirements of sliding window MAX is to remove a tuple with value v if there is another tuple in the window with a value greater than v and a younger timestamp. For example, in Figure 2.5, when the tuple with value 73 arrives, we can

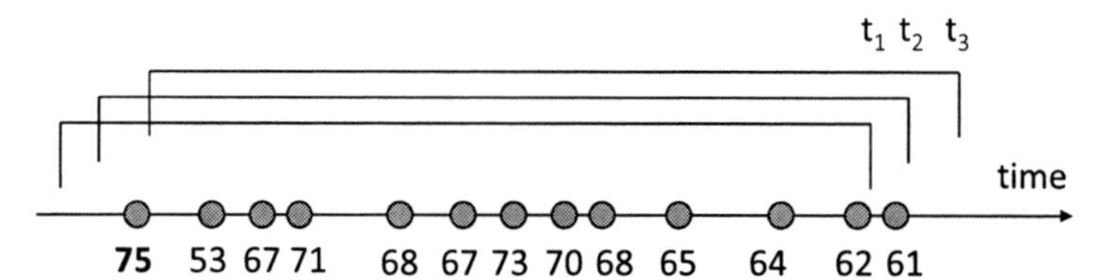

Figure 2.6: Finding the maximum element in a sliding window.

remove the five previous tuples (with values 53, 67, 71, 68 and 67) from the state. Similar kinds of optimizations have been proposed for more complex sliding window operators, such as skylines [Lin et al., 2005; Tao and Papadias, 2006] and top-k [Mouratidis et al., 2006].)

One downside of the negative tuple approach to maintaining a view of a continuous query is that twice as many tuples flow through the plan. While we cannot avoid negative tuples with operators such as negation, which are non-monotonic even on unbounded streams, we do not require them for monotonic operators, such as joins, over explicit time-based windows. The insight is to realize that in some cases, we can "predict" when tuples will expire. Thus, rather than generating negative tuples, operators can tag regular tuples with their expiration times. Applications can then maintain materialized views by removing tuples whose expiration times are smaller than the current time.

We formalize this observation as follows [Golab and Özsu, 2005]. As before, let $Q(\tau)$ be the result of a continuous query Q at time τ. For simplicity, assume that tuples arrive at the DSMS instantaneously and that results are produced instantaneously. Let $S(\tau)$ be the multi-set of tuples arriving on the input stream(s) of Q at time τ and let $S(0, \tau)$ be the multi-set of tuples that have arrived up to and including time τ. Next, let $P_S(\tau)$ be the new results of Q generated at time τ (i.e., the positive delta) and let $E_S(\tau)$ be results that were in $Q(\tau - 1)$ but are not in $Q(\tau)$ (i.e., the negative delta), given an input stream S. Let $P_S(0, \tau)$ and $E_S(0, \tau)$ be all the positive and negative deltas, respectively, up to and including time τ. In the query-as-view mode, $Q(\tau)$ evolves as follows.

$$\forall\tau \; Q(\tau + 1) = Q(\tau) \cup P_S(\tau + 1) - E_S(\tau + 1)$$

Now, continuous queries and their operators may be classified as follows.

- An operator is monotonic if $\forall\tau\forall S \; E_S(\tau) = \emptyset$. That is, results never "expire".

- An operator is *weakest non-monotonic* if $\forall\tau\forall S \; \exists c \in \mathbb{N}$ such that $E_S(\tau) = P_S(\tau - c)$. That is, irrespective of the input S, every result expires c time units after being generated.

- An operator is *weak non-monotonic* if $\forall\tau$ and $\forall S, S'$ such that $S(0, \tau) = S'(0, \tau)$, it is true that $\forall t \in P_S(0, \tau) \; \exists r$ such that $t \in E_S(r) \wedge t \in E_{S'}(r)$. That is, we can determine when a result tuple will expire at the time of its generation, but different result tuples may have different "lifetimes".

- An operator is *strict non-monotonic* if $\exists\tau\exists S, S'$ such that $S(0,\tau) = S'(0,\tau)$ and $\exists r \exists t \in P_S(0,\tau)$ such that $t \in E_S(r) \wedge t \notin E_{S'}(r)$. That is, at least some results have expiration times that depend upon future inputs and, therefore, cannot be predicted at generation time.

For example, selection over an explicit time-based window is weakest non-monotonic; a tuple with a timestamp t that satisfies the selection predicate will need to be removed from a materialized view of the output at time $t + w$, where w is the window length [Krämer and Seeger, 2009]. On the other hand, a join of two explicit time-based windows is weak non-monotonic. To see this, note that a join result expires when at least one of its input tuples expires. This creates results with various lifetimes, as a new tuple from one stream may join with a recent tuple from the other stream or one that is about to expire [Cammert et al., 2008]. Finally, note that queries over explicit count-based windows are strict non-monotonic since expirations from a count-based window are triggered by new tuples, whose arrival times are generally unpredictable. The significance of the above classification is that weakest and weak non-monotonic operators produce results with predictable expiration times and, therefore, do not need to produce explicit negative tuples.

Of course, even if we do not propagate negative tuples through a query plan, operators still need to purge expired tuples from their state, and, if applicable, generate new answers. For example, returning to Figure 2.6, the sliding window `MAX` operator may not be explicitly notified of expirations via negative tuples; instead, it needs to maintain the entire window and periodically check if the oldest tuple has expired. Similarly, the join operator needs to remove expired tuples from its state even if not explicitly signaled to do so by negative tuples.

2.5 STREAM QUERY OPTIMIZATION

Having discussed the basics of continuous query processing, we now present common DSMS query optimization strategies. As in a traditional DBMS, the objective is to find an efficient query plan. However, rather than minimizing the number of disk accesses, a continuous query plan should minimize the processing cost per unit time in order to keep up with the inputs [Kang et al., 2003]. Some techniques to achieve this are similar to those used in traditional DBMSs, such as reordering operators in a way that "filters out" as many tuples as possible early in the pipeline; others are specific to the resource-constrained, long-running nature of streaming queries. Independently, as discussed in Section 2.4.1, we may treat a DSMS as a real-time system and employ scheduling algorithms that optimize various quality-of-service metrics.

2.5.1 STATIC ANALYSIS AND QUERY REWRITING

We begin with a discussion of static analysis and optimization. When a DSMS receives a declarative query or a logical query plan, the first step is to ensure that the query can be evaluated in a non-blocking fashion using finite memory. Most DSMSs simply do not allow potentially "unsafe" queries such as joins of unbounded streams. However, some of these queries may, in fact, be executed using limited memory [Arasu et al., 2004a].

Consider two unbounded streams: $S(A, B, C)$ and $T(D, E)$. The query $\pi_A(\sigma_{A=D \wedge A>10 \wedge D<20}(S \times T))$ may be evaluated in bounded memory whether or not the projection preserves duplicates. To preserve duplicates, for each integer i between 11 and 19, we maintain the count of tuples in S such that $A = i$ and the count of tuples in T such that $D = i$. To remove duplicates, we store flags indicating which tuples have occurred such that $S.A = i$ and $T.D = i$ for $i \in [11, 19]$. Conversely, the query $\pi_A(\sigma_{A=D}(S \times T))$ is not computable in finite memory either with or without duplicates. Interestingly, $\pi_A(\sigma_{A>10}S)$ is computable in finite memory only if duplicates are preserved; in which case, we simply place each S-tuple with $A > 10$ on the output stream as soon as it arrives. On the other hand, the query $\pi_A(\sigma_{B<D \wedge A>10 \wedge A<20}(S \times T))$ is computable in bounded memory only if duplicates are removed: for each integer i between 11 and 19, we need to maintain the current minimum value of B among all the tuples in S such that $A = i$ and the current maximum value of D over all tuples in T.

Now, suppose that there is a logical plan for one or more queries. As in traditional DBMSs, we may try to rewrite the plan to improve efficiency. Many well-known rewriting rules apply, e.g., performing selections before joins and evaluating inexpensive predicates before complex ones [Babu et al., 2004a; Golab et al., 2008b]. New rules have also been proposed for novel continuous query operators. For example, selections and explicit time-based windows commute but not selections and explicit count-based windows [Arasu et al., 2006] (to see this, note that an explicit count-based window generates a negative tuple whenever a new tuple arrives; if we execute a selection operator first, it may drop some tuples and prevent the window operator from producing the required negative tuples).

Finally, some rewritings are possible if certain constraints can be guaranteed to hold on the input(s). For example, we can compute a join of unbounded streams using very little memory if we know that matching tuples, if any, arrive at most t time units apart [Babu et al., 2004b]. In some cases, we can eliminate sliding window joins altogether [Golab et al., 2008a].

2.5.2 OPERATOR OPTIMIZATION - JOIN

Next, we discuss optimizing physical implementation of various operators, beginning with the join.

First, recall that joins are often evaluated on implicit or explicit sliding windows. In both cases, the expired tuples need to be removed over time. One solution is to check for expired tuples after every time tick. However, this may be expensive and unnecessary. For instance, in a hash join, it may be necessary to check each hash bucket for expired tuples. Alternatively, when a new tuple arrives to the join operator and is inserted into its hash bucket (and then probes the corresponding hash bucket of the other window), the expired tuples can be removed from that particular bucket. Another choice is to perform expiration periodically; in which case, the join operator may encounter expired tuples during the probing stage and need to ignore them (expired tuples may be identified by comparing their timestamps to the current time) [Golab and Özsu, 2003b]. Thus, periodic expiration reduces the cleanup costs but increases the memory usage and join processing costs.

Another opportunity for optimization involves joins of multiple streams. Suppose that we need to join $S1$, $S2$ and $S3$ on a common attribute. When a new tuple arrives on $S1$, there are two choices: probe $S2$ first or probe $S3$ first. If $S2$ is expected to produce fewer matches than $S3$, $S2$ should be probed first to minimize the number of intermediate results [Golab and Özsu, 2003b; Viglas and Naughton, 2002; Viglas et al., 2003] and "short-circuit" the join operation as soon as possible [Yang and Papadias, 2008].

2.5.3 OPERATOR OPTIMIZATION - AGGREGATION

Recall from Section 2.4.3 that maintaining the result of a sliding window aggregate function at all times is difficult because it is necessary to react to every new and expired tuple. Re-computing the aggregate periodically by grouping on the time attribute is a more efficient and often more user-friendly solution (users may find it easier to deal with periodic output rather than a continuous output stream [Arasu and Widom, 2004; Chandrasekaran and Franklin, 2003]). Below, we describe several types of data structures called *synopses* that store enough state to efficiently re-compute various aggregates.

Figure 2.7 illustrates a *prefix synopsis* that stores pre-computed values over prefixes of the stream [Arasu and Widom, 2004]. This synopsis is suitable for *subtractable* aggregate functions such as `SUM` and `COUNT`. An aggregate f is subtractable if, for two multi-sets X and Y such that $X \supseteq Y$, $f(X - Y) = f(X) - f(Y)$ [Cohen, 2006]. A prefix synopsis is associated with three parameters: s, which is the time between updates (re-computations), b, which defines the longest window covered by the synopsis (i.e., the synopsis allows us to compute the aggregate over any window length up to b, so long as it is a multiple of s), and f, which is the type of aggregate function used to create the synopsis. Let t be the last update time. The synopsis stores aggregate values over b prefixes: $f([1,t]), f([1,t-s]), f([1,t-2s]), \dots, f([1,t-bs])$. To compute f over a window of size ns (where $n < b$) at time t, i.e., $f((t-ns,t])$, we compute $f([1,t]) - f([1,t-ns])$. The example shown in Figure 2.7 computes `SUM` over a window of size $7s$ using a prefix synopsis with $b = 8$ and $f =$ `SUM`. The next synopsis update, which takes place at time $t+s$, replaces $f([1,t-8s])$ with $f([1,t+s])$, where $f([1,t+s])$ can be computed as $f([1,t]) + f((t,t+s])$. This can be done efficiently by pre-computing $f((t,t+s])$ incrementally as new tuples arrive.

An *interval synopsis* is used with distributive aggregates (recall Section 2.2.2) that are not subtractable, such as `MIN` and `MAX` [Arasu and Widom, 2004; Bulut and Singh, 2003, 2005; Zhang and Shasha, 2006; Zhu and Shasha, 2003]. An interval synopsis with parameters s, b, and f (as defined for the prefix synopsis; additionally, assumes that b is a power of two) stores the values of f over $2b$ intervals. In particular, there are b disjoint intervals of length s, $\frac{b}{2}$ disjoint intervals of length $2s$, and so on up to one interval of length bs. Figure 2.8 shows an interval synopsis with $b = 8$ and $f =$ `MAX` to compute `MAX` over a window of size $7s$. To do this, we take the maximum of the values stored in three disjoint intervals; in general, we need to access $\log b$ intervals. During the next update at time $t+s$, we can drop $f((t-8s,t-7s])$ because this interval now references expired tuples. Furthermore, we insert $f((t,t+s])$. As before, we can incrementally pre-compute

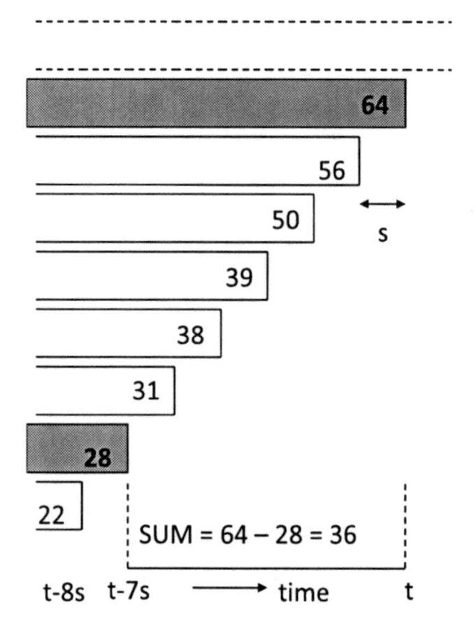

Figure 2.7: A prefix synopsis.

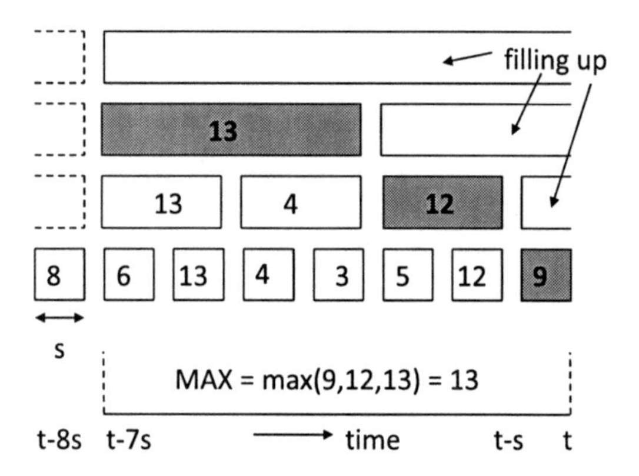

Figure 2.8: An interval synopsis.

$f((t, t+s])$ as new tuples arrive. Moreover, interval $(t-s, t+s]$ is now full, and its value may be computed as $\max(f((t-s, t]), f((t, t+s]))$.

Note that algebraic aggregates may be computed using the synopses of appropriate distributive aggregates. For instance, since `AVG = SUM/COUNT`, a query computing the average may use `SUM` and `COUNT` synopses.

Holistic aggregates may use an interval synopsis if it stores additional information per interval. For instance, we can answer `QUANTILE`, `TOP k`, and `COUNT DISTINCT` queries by storing the full frequency histogram in each interval. Alternatively, we can store a sketch (recall Section 2.2.2) in

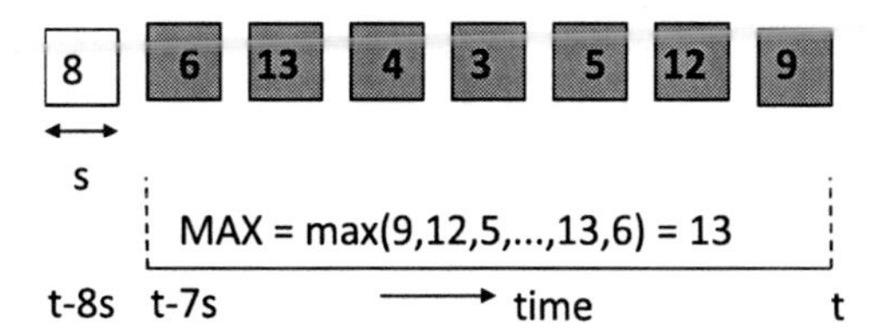

Figure 2.9: A basic interval synopsis.

each interval to obtain approximate answers [Arasu and Manku, 2004; Lee and Ting, 2006]. Of course, this is possible only if we can merge sketches of non-overlapping intervals to produce a new sketch. For example, we can merge two CM sketches simply by adding up the corresponding entries in their arrays.

Technically, we may also store frequency counts or sketches in a prefix synopsis since many sketches are subtractable. However, each interval of a prefix synopsis summarizes the entire stream up to a certain point. Therefore, the synopsis may store counts of values that appeared a long time ago and may never be seen again. This leads to excessive space usage and increased computation costs. Additionally, given two sketches of the same size, one summarizing the interval $[1, t+s]$ and another summarizing $(t, t+s]$, the latter should be more accurate since it does not have to approximate the entire distribution of the stream.

We can reduce the space usage and update time of an interval synopsis by storing only the b short intervals "at the bottom" [Golab et al., 2004; Krishnamurthy et al., 2006; Li et al., 2005b; Zhu and Shasha, 2002]. The tradeoff is that querying such a synopsis is not as efficient as before: we need to probe up to b intervals to compute the answer, up from $\log b$ intervals. We refer to this modified synopsis as a *basic interval synopsis*. Figure 2.9 gives an example for $b = 8$ and a `MAX` query over a window of length $7s$.

Finally, we note that there exist synopses for approximating the current value of an aggregate function at any time with bounded error. In contrast to basic interval synopses that maintain intervals of equal size, the idea is to construct small intervals for recent data and longer intervals for older data. This data structure is referred to as an *Exponential Histogram* and may be used for sliding window aggregates such as sum [Cohen and Strauss, 2003; Datar et al., 2002; Gibbons and Tirthapura, 2002], variance and k-medians clustering [Babcock et al., 2003], histograms [Qiao et al., 2003], and order statistics [Lin et al., 2004; Xu et al., 2004].

2.5.4 MULTI-QUERY OPTIMIZATION

As we discussed above, aggregate queries over different window lengths and possibly different `SLIDE` intervals [Golab et al., 2006b; Salehi et al., 2009] may share state and data structures. Similarly, we can share state and computation across similar predicates and joins [Denny and Franklin, 2005; Dobra et al., 2002; Hammad et al., 2003; Hong et al., 2009; Zhang et al., 2005]. In general, a DSMS may group similar queries and run a single query plan per group [Chen et al., 2000, 2002;

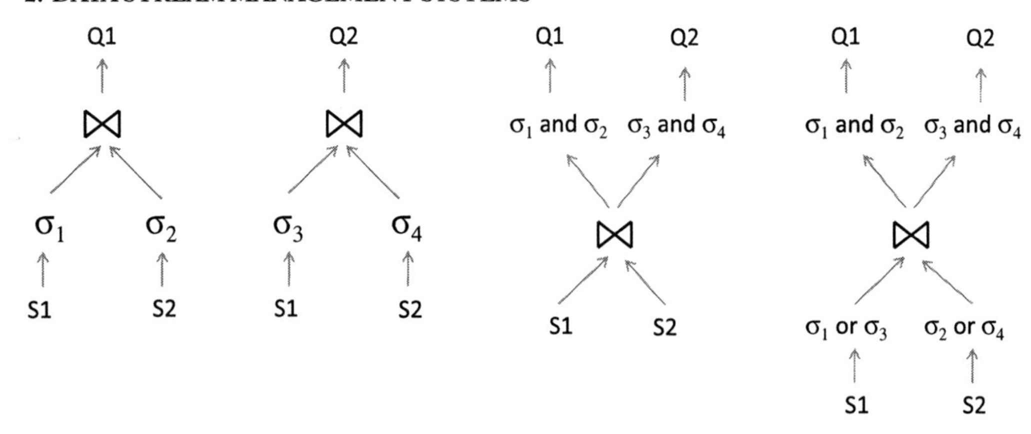

Figure 2.10: Separate and shared query plans for $Q1$ and $Q2$.

Krishnamurthy et al., 2004; Wang et al., 2006]. Figure 2.10 shows some of the issues involved in shared query plans. The first two plans correspond to executing queries $Q1$ and $Q2$ separately, in which selections are evaluated before joins. The third plan executes both queries and evaluates the join first, then the selections (note that the join operator effectively creates two copies of its output stream). Despite sharing work across two queries, the third plan may be less efficient than separate execution if only a small fraction of the join result satisfies the selection predicates σ_1 through σ_4. If so, then the join operator will perform a great deal of unnecessary work over time. The fourth plan addresses this problem by "prefiltering" [Golab et al., 2008b] the streams before they are joined.

For workloads consisting of a large number of queries with similar selection predicates, predicate indexing is a useful optimization technique [Chandrasekaran and Franklin, 2003; Demers et al., 2006; Fabret et al., 2001; Hanson et al., 1999; Krishnamurthy et al., 2006; Lim et al., 2006; Madden et al., 2002b; Wu et al., 2004]. The idea is to build an index, such as a balanced tree, for each attribute. When a new tuple arrives, we probe the predicate indices with the tuple's attribute values to find all the queries that are satisfied by this tuple.

2.5.5 LOAD SHEDDING AND APPROXIMATION

In some situations, a DSMS may not be able to evaluate all queries on every new tuple. One general solution, which works particularly well for aggregates, is to report new answers less often or produce approximate answers [Denny and Franklin, 2006], as already discussed. If this is not sufficient, the DSMS may be forced to shed load by dropping a fraction of the input. The simplest load shedding strategy is to randomly sample the input. Alternatively, *semantic load shedding* exploits properties of the stream and/or the queries to drop "less important" tuples [Gedik et al., 2008; Tatbul et al., 2003].

For an example of semantic load shedding, suppose that we are executing a sliding window join, and we cannot keep up with the inputs or we do not have enough memory to store the windows.

Suppose that our goal is to produce as many result tuples as we can. The idea is that tuples that are about to expire or tuples that are not expected to produce many join results should be dropped (in case of memory limitations [Das et al., 2005; Li et al., 2006a; Xie et al., 2005]), or inserted into the join state but ignored during the probing step (in case of CPU limitations [Ayad et al., 2006; Gedik et al., 2005]). Note that other objectives may be more appropriate in some cases, such as obtaining a random sample of the join result [Srivastava and Widom, 2004b].

In general, if a DSMS needs to shed load, it should do so in a way that minimizes the drop in accuracy. This problem becomes more difficult when multiple queries with many operators are involved, as we must decide where in the query plan the tuples should be dropped. Clearly, dropping tuples early in the plan is the most effective, but this affects the accuracy of many queries if the plan is shared. On the other hand, load shedding later in the plan, after the shared sub-plans have been evaluated and the only remaining operators are specific to individual queries, may have little effect in reducing the system load. This gives rise to the problem of optimal placement of sampling operators in multi-query plans. There are known results for random load shedding for windowed aggregates [Babcock et al., 2004b] and for quality-of-service-driven load shedding for windowed aggregates [Tatbul and Zdonik, 2006].

Another issue in this context is to decide how much load to shed. One possible approach is to monitor the sizes of operator queues [Tu et al., 2006].

Finally, we note that it is not clear if an optimal query plan chosen without load shedding is still optimal if load shedding is used. This turns out to be true for sliding window aggregates but not for sliding window joins [Ayad and Naughton, 2004].

2.5.6 LOAD BALANCING

Instead of shedding load, it may be possible for a DSMS to defer it. For example, in some bursty environments, it may be feasible to write some tuples to disk during overload and process them later during quiet periods [Liu et al., 2006a; Reiss and Hellerstein, 2005].

Interestingly, some continuous queries may cause self-inflicted overload. For example, recall query $Q1$, which computes the total traffic originating from each source IP address every minute. A straightforward way of evaluating such a group-by-aggregation query is to immediately output the new aggregate value for each group at the end of each tumbling window. Another approach, which ensures a more balanced load at the expense of higher latency, is to output a new value for group g only after we have seen at least one tuple belonging to g in the next window. In the GS Tool, this approach is referred to as *slow flush* [Johnson et al., 2005c].

2.5.7 ADAPTIVE QUERY OPTIMIZATION

The per-unit-time cost of a continuous query plan may change over time for three reasons: 1) change in the processing time of an operator due to a change in system conditions such as available CPU or memory, 2) change in the selectivity of a predicate, and 3) change in the arrival rate of a stream [Avnur and Hellerstein, 2000]. A DSMS may, therefore, change query plans on-the-fly, which may require transforming and migrating the state of operators in the old plan to operators in the new plan [Babu and Widom, 2004; Babu et al., 2005; Deshpande and Hellerstein, 2004; Zhu et al., 2004]. Alternatively, instead of using a fixed query plan at any point in time, the *Eddies* approach routes each tuple through all the query operators in some order that is independent of the evaluation order of other tuples [Avnur and Hellerstein, 2000; Bizarro et al., 2005; Chandrasekaran and Franklin, 2003; Deshpande, 2004; Gu et al., 2007; Madden et al., 2002b; Raman et al., 2003; Tok and Bressan, 2002]. In effect, the query plan is dynamically re-ordered to match the current system conditions. This is accomplished by tuple routing policies that attempt to discover which operators are fast and selective, and those operators are scheduled first. A recent extension adds queue length as the third factor for tuple routing strategies in the presence of multiple distributed Eddies [Tian and DeWitt, 2003]. There is, however, an important trade-off between the resulting adaptivity and the overhead required to route each tuple separately.

2.5.8 DISTRIBUTED QUERY OPTIMIZATION

Distributing a DSMS across a cluster of machines is necessary to deal with high-speed data streams. There are two broad issues in distributed DSMS optimization: parallelizing and distributing the system itself, and shifting some computation to the sources of the data. We discuss each issue below.

In terms of distributing the DSMS, there are at least two options: we can split the query plan across multiple processing nodes, or we can partition the stream and let each node process a subset of the data to completion [Abadi et al., 2005; Johnson et al., 2008]. Partitioning the query plan involves assigning query operators to nodes [Xing et al., 2006] and may require re-balancing over time [Liu et al., 2005; Shah et al., 2003; Xing et al., 2005]. Partitioning the data is usually done once for a given query workload. The simplest stream partitioning method is to route incoming tuples to processing nodes in a round robin fashion. However, this approach is suboptimal for may classes of continuous queries, including joins and aggregates with grouping. For example, suppose that we want to evaluate query $Q1$ from Section 2.1.1 in a distributed fashion because the incoming stream is too fast to be processed by a single machine. Round-robin partitioning may assign tuples with the same source IP address to different nodes. This means that each node may produce partial sums for a number of groups, which will have to be combined to obtain the final answer at the end of each window. On the other hand, if we split the stream by source IP address, each node will have enough information to compute the final aggregate for some subset of IP addresses. Hence, hash-based partitioning on the group-by attributes is more efficient.

Query processing at the sources, also known as *in-network processing*, can reduce the communication overhead between the sources and the DSMS, and the load on the DSMS. Distributive aggregates over tumbling windows can be easily computed in this fashion—rather than sending all the "raw" data to the DSMS, each source can pre-aggregate its data and send partial aggregates. The DSMS then combines the partial aggregates into a final answer. Holistic aggregates, on the other hand, cannot be pre-computed by the sources and merged at the DSMS. However, we can obtain approximate answers to holistic aggregates using in-network processing by having each source pre-compute a sketch. The DSMS can then combine the sketches and compute the final answer. This procedure is very similar to merging sketches in a sliding window synopsis (recall Section 2.5.3).

There exist further optimization opportunities if the sources can communicate with each other in addition to communicating directly with the DSMS [Madden et al., 2002a; Nath et al., 2004]; for example, the communication network of the sources and the DSMS may be a tree (with the DSMS at the root). For distributive aggregates, rather than having each source send pre-aggregated data directly to the DSMS, the sources can merge their pre-aggregated data with those of their descendants.

The above techniques apply to queries that produce a time-evolving output stream, such as aggregates over tumbling windows. Now, suppose that we want to maintain the latest result of some function that requires input from a set of distributed sources. For instance, we may want to keep track of the average speed of a set of vehicles. In a naive solution, each source generates a continuous stream of updates to its current speed. A more efficient solution is to keep track of an approximate answer (with bounded error) by requesting updates from the sources, only if the current answer exceeds the specified error bound [Babcock and Olston, 2003; Cheng et al., 2005; Cormode and Garofalakis, 2005; Cormode et al., 2005, 2006; Das et al., 2004; Jain et al., 2004; Keralapura et al., 2006; Manjhi et al., 2005; Olston et al., 2003; Sharfman et al., 2007].

CHAPTER 3

Streaming Data Warehouses

This chapter covers the design of Streaming Data Warehouses (SDWs). In some ways, an SDW faces the same challenges as standard data warehouses, among them the need to store massive amounts of data on disk for off-line analysis. However, SDWs must also deal with DSMS-like issues such as reacting to continuously arriving data.

Recall the abstract system architecture from Figure 1.4. In order to load new data as quickly as possible, it is essential to ensure the efficiency of the ETL process (for base tables) and the update scheduling and propagation mechanism (for derived tables). We will discuss how to do so below, along with some novel query processing issues that arise in an SDW.

3.1 DATA EXTRACTION, TRANSFORMATION AND LOADING

When new data arrive at the warehouse, they may need to be transformed before being loaded into their base tables. Simple ETL tasks such as unzipping compressed files and standardizing attribute values are easy to perform in a streaming fashion. However, some parts of the ETL process may be more complex. For instance, a common transformation joins new data with a disk-resident table R that contains various descriptive attributes, such as human-readable router names corresponding to numerical router IDs found in a network performance feed. In a traditional warehouse that performs updates during downtimes, we can buffer new data in a temporary table and then join the entire temporary table with R using a standard join operator. In an SDW, we need to perform the join incrementally as new data arrive. We have already discussed joining streams with relations in the previous chapter. However, in this context, the relation is on disk and must be paged into main memory, which introduces new challenges. A simple solution is to build an index on R and probe it for every new tuple. However, this random access of R may cost up to one disk I/O per new tuple. Instead, an algorithm called *Mesh Join* has been proposed that continually scans R back and forth in a sequential fashion, and it brings in one page at a time into memory [Polyzotis et al., 2008]. New tuples are buffered only until they have seen all the pages of R. When a new tuple has joined with all of R, it can leave the buffer and is ready to be loaded into its base table. The next new tuple then enters the buffer and begins probing the part of R that is currently in memory. Note that the Mesh Join does not require R to be indexed and allows one-to-many joins from the new data to R.

After a batch of tuples has gone through the required transformations, it can be loaded into the corresponding base table. Base and derived tables in an SDW are usually partitioned by a timestamp attribute so that insertions of new data only affect a small number of recent partitions

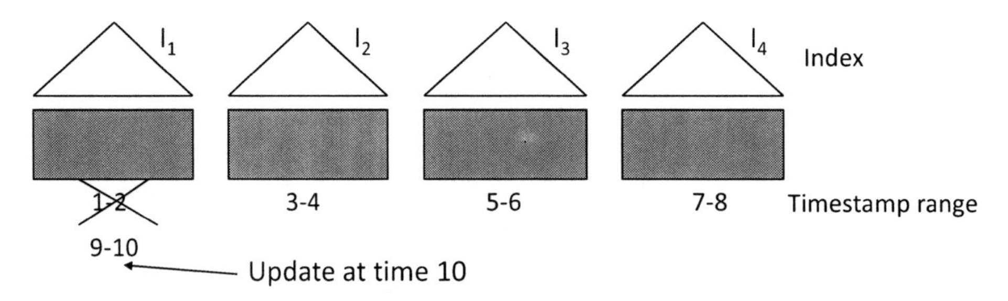

Figure 3.1: Partitioning a table on a timestamp attribute.

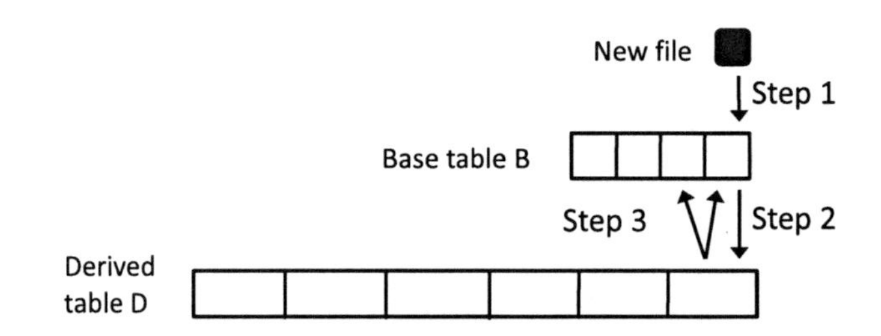

Figure 3.2: Updating a partitioned derived table.

[Folkert et al., 2005; Golab et al., 2009a; Shivakumar and García-Molina, 1997]. Figure 3.1 shows a simple example with four separately indexed partitions of size two each. At time eight, all four partitions are full and contain tuples with timestamps in the specified range. At time ten, we replace the oldest partition with new data. If tuples arrive in timestamp order, then each update requires disk access to only one partition.

3.2 UPDATE PROPAGATION

An efficient ETL process ensures that new data are loaded into base tables as soon as possible. The next step is to efficiently propagate changes across multiple layers of derived tables (materialized views). Again, this problem is simpler in traditional warehouses that are updated during downtimes—a brute force solution is to recompute all the views or at least those which cannot be maintained incrementally. In an SDW, the idea is to avoid recomputing an entire derived table and, instead, recompute only those partitions that have changed. In Figure 3.2, we show a new file being loaded into (the most recent partition of) a base table B (Step 1). If we know the partition dependencies between B and the derived table D, as shown, then we can infer that only the most recent partition of D needs to be updated (Step 2). Furthermore, if we know the partition dependencies in the other direction, from D to B, we can infer that recomputing the most recent D-partition requires reading data from the two most recent B-partitions (Step 3).

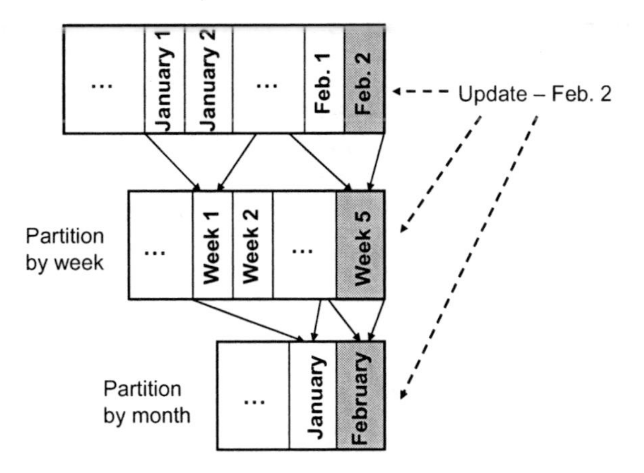

Figure 3.3: Identifying out-of-date partitions of rolled-up tables in response to an update to a fact table.

The challenge in efficient update propagation is to identify which partitions of a derived table may have changed after a base table has been updated. In some cases, we can use dimension hierarchies to find partitions that need updating. Figure 3.3 shows that if daily sales data are added to a fact table, then we can determine which partitions of the weekly and monthly roll-up tables are out-of-date.

In general, partition dependencies may not be obvious from the SQL specification of a derived table and may need to be explicitly stated by the user. For example, the DataDepot warehouse [Golab et al., 2009a] requires each view definition to include `SOURCE_LOWER_BOUNDS` and `SOURCE_UPPER_BOUNDS` values that determine the relationship between partitions of a derived table and partitions of its source(s). Suppose that derived table D is sourced from a single base table B having 8 partitions of length one day each. Figure 3.4(a) illustrates the simplest case, in which D also consists of 8 partitions of one day each (in practice, D may have many more partitions than B). The upper and lower bounds are the same; in other words, the Pth partition of S is derived from the Pth partition of B. In Figure 3.4(b), D stores two days of data per partition, making the lower and upper bounds $P * 2$ and $P * 2 + 1$, respectively. That is, two consecutive source partitions of one day each are needed to compute one derived partition; the multiplication factor signifies a change of scale from one-day to two-day partitions. This example is similar to a traditional roll-up table. In Figure 3.4(c), D is a sliding window aggregate extending four days into the past, as specified by the lower and upper bounds of $P - 3$ and P, respectively.

We note that partition dependencies can be thought of as tracing the *lineage* of a table and its partitions. Coarse-grained (i.e., partition-based) lineage is sufficient for an SWD to ensure efficient maintenance of derived tables, in contrast to tuple-based lineage tracing used in data exploration [Cui and Widom, 2000], probabilistic data management [Widom, 2005], and multi-query optimization [Krishnamurthy et al., 2006; Madden et al., 2002b].

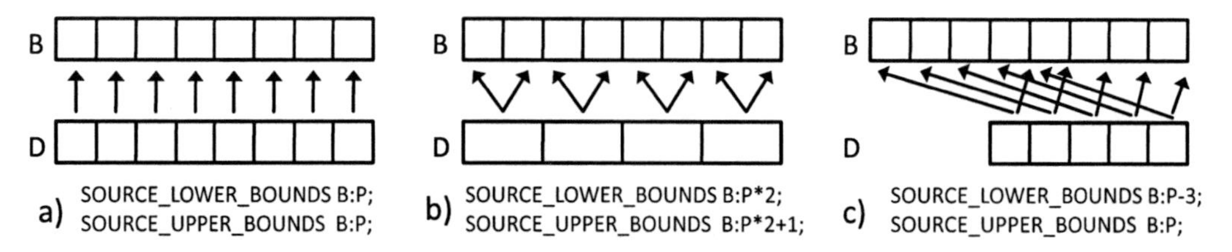

Figure 3.4: Examples of partition dependencies.

3.3 DATA EXPIRATION

As shown in Figure 3.1, an SDW maintains temporally partitioned tables and replaces old partitions with new ones over time. These are similar to sliding windows maintained by a DSMS, but they are usually much longer; some derived tables may store months or even years of historical data. So far, we have used a single timestamp for partitioning and expiration, which implies that every tuple in a warehouse table has the same lifetime. However, as we discussed in Section 2.4.3, weak non-monotonic operators such as joins over explicit windows generate results with variable lifetimes.

Consider storing tuples with variable lifetimes in the partitioned table from Figure 3.1. At time ten, we will insert new data into partition I_1, as before. However, we may have to remove expired tuples from each partition. To see this, note that it is no longer true that only tuples that were inserted at times 1 and 2 will expire at time 10. Similarly, if we partition the table on the *expiration* timestamp rather than the generation timestamp, we can limit expirations to a single partition, but we may have to insert new tuples into every partition. This defeats the purpose of partitioning.

A better solution for storing tuples with variable lifetimes is a doubly-partitioned table on the insertion and expiration timestamps [Golab et al., 2006c], an example of which is illustrated in Figure 3.5. As in our previous example, there are four partitions that are updated every two time units; however, tuples now have lifetimes between one and eight time units. Furthermore, the four partitions are now created by dividing the insertion and expiration times into two ranges each. As illustrated at the top of Figure 3.5, at time 8, partition I_1 stores tuples that have arrived between times one and four that will expire between times 9 and 12 (the other three partitions may be described in a similar way). The update illustrated on the bottom of Figure 3.5 occurs at time ten, inserts new tuples into I_1 and I_2, and deletes expired tuples from I_1 and I_3. Observe that I_4 does not have to be accessed during this update or during the next update at time 12. Then, the next two updates at times 14 and 16 will insert into I_3 and I_4, and delete from I_2 and I_4 (I_1 will not be accessed). In general, increasing the number of partitions leads to more partitions not being accessed during updates.

One problem with the doubly-partitioned table in Figure 3.5 is that its partitions may not have equal sizes. For example, at time 8, I_2 stores items that arrived between times one and 4 and will expire between times 13 and 16. That is, I_2 is empty at that time because there are no items

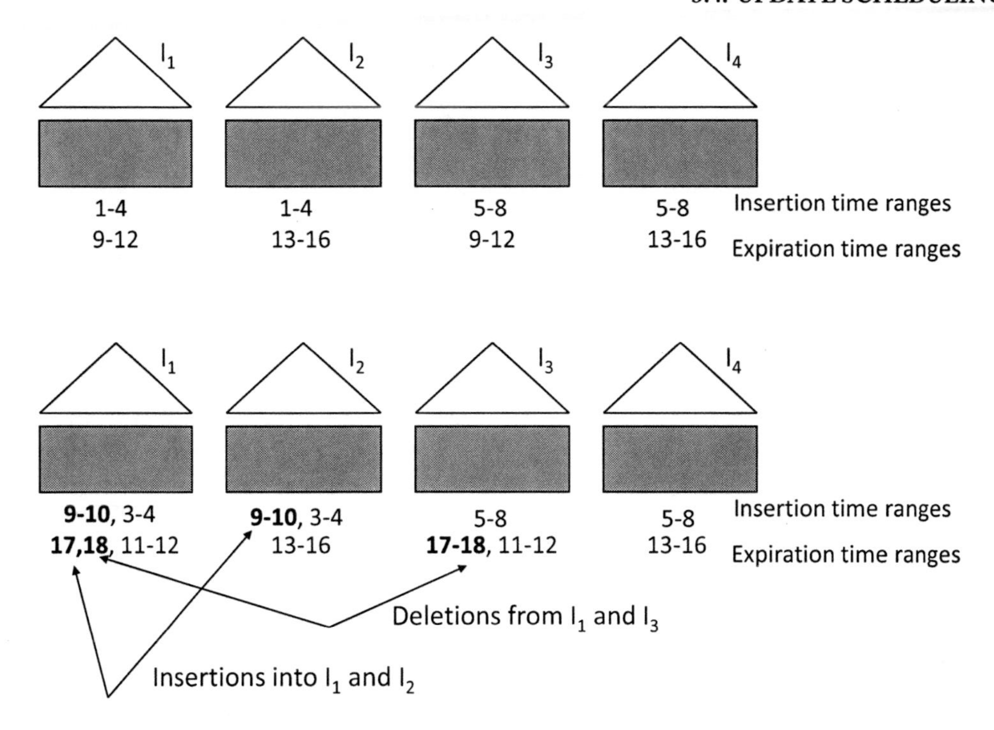

Figure 3.5: Partitioning a table on two attributes: insertion and expiration timestamp.

whose lifetimes are larger than 8. As a result, the other three partitions are large and their update costs may dominate the overall maintenance cost.

This problem can be addressed by changing the insertion and expiration times assigned to each partition. An improved technique is shown in Figure 3.6 for the same parameters as before (tuple have lifetimes of up to 8 time units and updates occur every two time units). Rather than dividing the insertion and expiration time ranges chronologically, we now distribute updates in round-robin fashion such that no partition incurs two consecutive insertions or expirations. For instance, the update illustrated in Figure 3.6 takes place at time 10, inserts new tuples into I_1 and I_2, and expires tuples from I_1 and I_3. The next update at time 12 inserts new tuples into I_3 and I_4, and deletes old tuples from I_2 and I_4. Spreading consecutive updates over different partitions ensures that partitions have similar sizes [Golab et al., 2006c].

3.4 UPDATE SCHEDULING

Having explained *how* an SDW propagates new data across derived tables, we now deal with the issue of *when* to perform updates. In practice, external sources push new data into the warehouse, so it is not feasible to request new data at a particular time of our choosing. Instead, we need to react to new data. Ideally, we want to load a batch of data as soon as it arrives and update all the affected

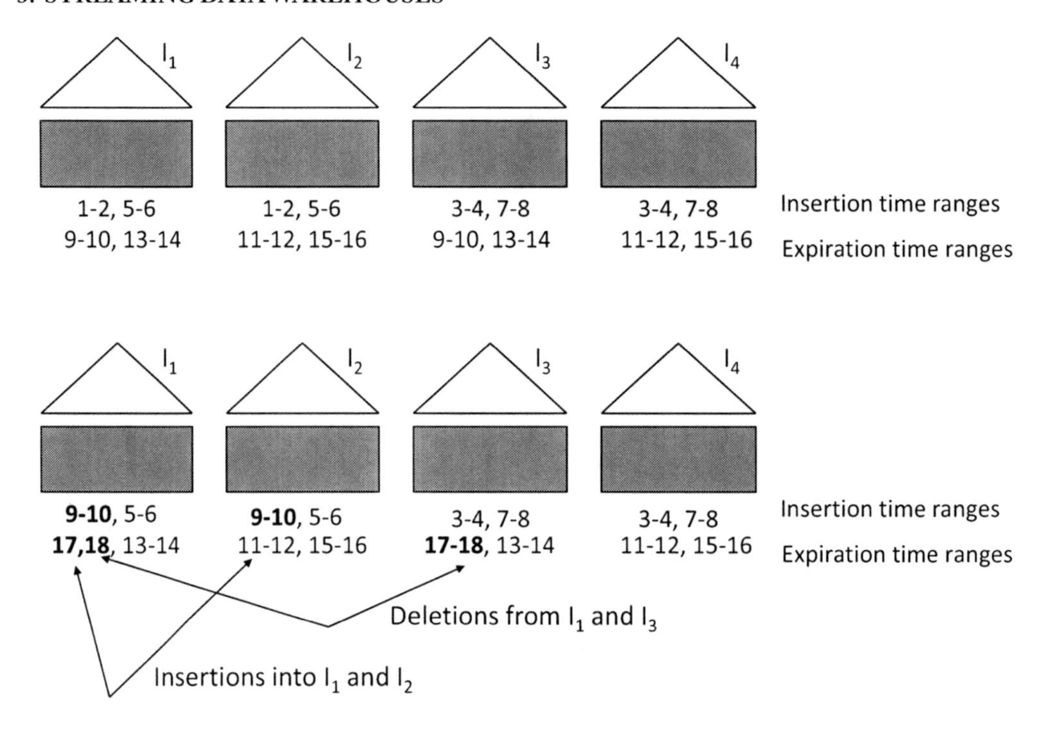

Figure 3.6: Another way of partitioning a table on the insertion and expiration timestamps.

derived tables. However, in practice, an SDW may receive tens or hundreds of data feeds and may maintain hundreds of derived tables. Without a limit on the number of tables being updated in parallel, we risk memory, CPU and disk arm thrashing, which severely impacts performance. One way to control resource usage is by using a *scheduler* to invoke table updates [Bateni et al., 2009; Golab et al., 2009b].

A scheduler requires an optimization metric to guide its decisions. Real-time systems typically aim to schedule jobs before their deadlines. In contrast, a natural metric for an SWD is to minimize data *staleness*. There are several reasonable ways to define staleness [Adelberg et al., 1995; Cho and Garcia-Molina, 2000; Labrinidis and Roussopoulos, 2001], one of which is to take the difference between the current time and the most recent tuple loaded into a given table. If tables have priorities, then minimizing priority-weighted staleness is more appropriate.

Figure 3.7 plots the staleness of a streaming warehouse table over time. Suppose that the first batch of new data arrives at time 4 with tuples that have been produced up to time 3 (e.g., it may have taken one unit of time for the data to arrive). Staleness accrues linearly until the data have been loaded at time 5. At that time, staleness drops to two since data up to time 3 have been loaded. Suppose that the next batch of data arrives at time 7, with data up to time 7, and that the third batch arrives at time 9, with data up to time 9. Suppose that the SDW loads both batches at time 11 (e.g.,

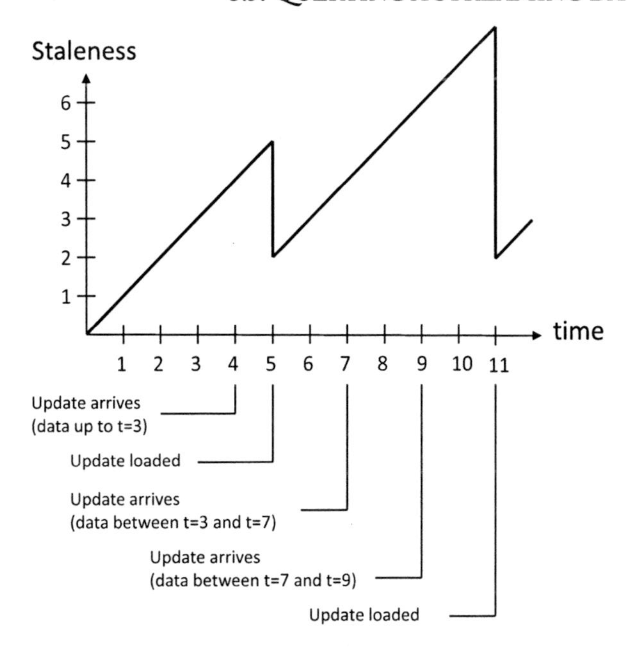

Figure 3.7: A plot of the staleness of a SDW table over time.

it may have been busy loading other tables and could not process the second batch right away). In this case, staleness accrues linearly between time 5 and time 11, and drops to 2 at time 11 since data up to time 9 have been loaded. Note that if the second batch of data did not arrive, and instead the third batch arrived at time 9 with data generated between times 3 and 9, then the staleness function would be exactly the same.

A simple greedy heuristic for minimizing priority-weighted staleness is as follows [Golab et al., 2009b]. Of all the tables that are currently out-of-date, we update the one whose priority-weighted staleness can be improved by the greatest amount per unit of processing time. Note that computing the possible improvement in staleness requires the knowledge of view hierarchies. If a derived table D is stale, and so are its sources, then attempting to update it gives a staleness improvement of zero (we must update the sources first).

3.5 QUERYING A STREAMING DATA WAREHOUSE

Having discussed techniques for efficiently maintaining data in an SDW, we now turn to query processing. The first issue is the overhead of partitioned tables. Suppose that table D receives new data every 15 minutes and that we want to maintain a 2-year historical window. From the point of view of updates, we can avoid accessing existing data by maintaining 15-minute partitions (each update will be loaded into its own partition). However, this requires $4 * 24 * 365 = 35040$ partitions in total, and extracting data from a particular day requires accessing 96 separate partitions. On the other hand, daily partitions may be easier to manage, but they may have to be repeatedly re-computed as new data arrive throughout the day. This may be solved by creating variable-sized partitions. For example, we can partition the most recent data into 48 partitions of size 15 minutes each, and partition the rest of the data by day [Golab et al., 2009b]. Of course, over time, the SDW needs to *roll-up* the 15-minute partitions into daily partitions.

The second problem involves data availability and concurrency control. Given that tables are updated frequently, we want to ensure that queries are not "blocked" by updates, and that queries read consistent data even if an update occurs while a query is running. One solution is to use multi-version concurrency control at the partition level [Golab et al., 2009b]. That is, for each table, we maintain a *partition directory* with pointers to the current partitions. When updating a table, we create new copies of all the partitions that have changed and construct a temporary partition directory that points to the new copies. At the end of the update, we replace the old partition directory with the new one and garbage-collect old partitions (provided that they are not being used by a query). When a query is issued, it copies the current partition directory and uses it until it has read all the data. An alternative solution is to take the latest update into account, provided that the query has not yet accessed any partitions that were affected by the update [Golab et al., 2006a].

Finally, updating tables as soon as new data arrive may cause queries to read "unstable" partitions even with multi-version concurrency control. For example, suppose that a base table B is partitioned by hour and receives new data every 15 minutes. If a new update arrives at the beginning of an hour and starts a new partition, this partition will be updated three more times within the next hour. Thus, running the same query on B at different times throughout the hour may give different answers. A simple way to ensure more "stability" in the result is to omit partitions that are not yet full when evaluating a query. To determine if a partition is full, we recursively check if all of its source partitions are full; assuming that data arrive in timestamp order, a base table partition spanning timestamps t_1 through t_2 is full if at least one tuple with a timestamp greater than t_3 has arrived. Of course, some queries may require access to all the available data, so an SDW should be able to return "stable" or "unstable" results.

CHAPTER 4

Conclusions

In this lecture, we discussed end-to-end stream data management, including Data Stream Management Systems for on-line query processing and Streaming Data Warehouses for off-line analysis. We conclude with some directions for future work.

Chapter 2 focused on systems that extend the relational model and support relational-like queries with selections, joins and aggregations. There has been some recent work on supporting other types of queries on streaming data, such as pattern matching and event processing. Incorporating these new operators in a general-purpose DSMS is an important practical problem.

As the amount of streaming data increases, DSMSs need to become even more scalable than they are today. One solution is to take advantage of new hardware. For example, in network management, we can perform some low-level data processing on routers and network interface cards using fast content-addressable memories.

Streaming Data Warehouses have appeared very recently and will likely be improved over time. Many aspects of SDWs require further research, including update scheduling strategies for optimizing various objectives, and monitoring data consistency and quality as new data arrive.

There are a number of relevant issues that this lecture has not covered or has covered briefly. Mining of data streams is one important topic that is not included. Another issue that is not included is fault-tolerance. We also discuss the resource management and distributed processing issues only briefly.

Bibliography

D. Abadi, D. Carney, U. Cetintemel, M. Cherniack, C. Convey, S. Lee, M. Stonebraker, N. Tatbul, and S. Zdonik. Aurora: A new model and architecture for data stream management. *VLDB J.*, 12 (2):120–139, Aug 2003. DOI: 10.1007/s00778-003-0095-z 6, 9, 12, 16

D. Abadi, Y. Ahmad, M. Balazinska, U. Cetintemel, M. Cherniack, J-H. Hwang, W. Lindner, A. Rasin, N. Tatbul, Y. Xing, and S. Zdonik. The design of the Borealis stream processing engine. In *Proc. 2nd Biennial Conf. on Innovative Data Systems Research*, pages 277–289, 2005. 6, 36

B. Adelberg, H. Garcia-Molina, , and B. Kao. Applying update streams in a soft real-time database system. In *Proc. ACM SIGMOD Int. Conf. on Management of Data*, pages 245–256, 1995. DOI: 10.1145/568271.223842 44

C. Aggarwal, editor. *Data Streams: Models and Algorithms*. Springer, 2007. 6

J. Agrawal, Y. Diao, D. Gyllstrom, and N. Immerman. Efficient pattern matching over event streams. In *Proc. ACM SIGMOD Int. Conf. on Management of Data*, pages 147–160, 2008. DOI: 10.1145/1376616.1376634 6

N. Alon, Y. Matias, and M. Szegedy. The space complexity of approximating the frequency moments. In *Proc. 28th Annual ACM Symp. on Theory of Computing*, pages 20–29, 1996. DOI: 10.1145/237814.237823 16

A. Arasu and G. S. Manku. Approximate counts and quantiles over sliding windows. In *Proc. ACM SIGACT-SIGMOD Symp. on Principles of Database Systems*, pages 286–296, 2004. DOI: 10.1145/1055558.1055598 33

A. Arasu and J. Widom. Resource sharing in continuous sliding-window aggregates. In *Proc. 30th Int. Conf. on Very Large Data Bases*, pages 336–347, 2004. 31

A. Arasu, B. Babcock, S. Babu, J. McAlister, and J. Widom. Characterizing memory requirements for queries over continuous data streams. *ACM Trans. Database Syst.*, 29(1):162–194, March 2004a. DOI: 10.1145/974750.974756 29

A. Arasu, M. Cherniack, E. Galvez, D. Maier, A. S. Maskey, E. Ryvkina, M. Stonebraker, and R. Tibbets. Linear road: a stream data management benchmark. In *Proc. 30th Int. Conf. on Very Large Data Bases*, pages 480–491, 2004b. 6

A. Arasu, S. Babu, and J. Widom. The CQL continuous query language: Semantic foundations and query execution. *VLDB J.*, 15(2):121–142, 2006. DOI: 10.1007/s00778-004-0147-z 7, 9, 11, 12, 17, 19, 30

R. Avnur and J. Hellerstein. Eddies: Continuously adaptive query processing. In *Proc. ACM SIGMOD Int. Conf. on Management of Data*, pages 261–272, 2000. DOI: 10.1145/342009.335420 35, 36

A. Ayad and J. Naughton. Static optimization of conjunctive queries with sliding windows over unbounded streaming information sources. In *Proc. ACM SIGMOD Int. Conf. on Management of Data*, pages 419–430, 2004. DOI: 10.1145/1007568.1007616 35

A. Ayad, J. Naughton, S. Wright, and U. Srivastava. Approximate streaming window joins under CPU limitations. In *Proc. 22nd Int. Conf. on Data Engineering*, page 142, 2006. 35

B. Babcock and C. Olston. Distributed top-k monitoring. In *Proc. ACM SIGMOD Int. Conf. on Management of Data*, pages 28–39, 2003. 37

B. Babcock, S. Babu, M. Datar, R. Motwani, and J. Widom. Models and issues in data streams. In *Proc. ACM SIGACT-SIGMOD Symp. on Principles of Database Systems*, pages 1–16, 2002. DOI: 10.1145/543613.543615 6, 7

B. Babcock, M. Datar, R. Motwani, and L. O'Callaghan. Maintaining variance and k-medians over data stream windows. In *Proc. ACM SIGACT-SIGMOD Symp. on Principles of Database Systems*, pages 234–243, 2003. 33

B. Babcock, S. Babu, M. Datar, R. Motwani, and D. Thomas. Operator scheduling in data stream systems. *VLDB J.*, 13(4):333–353, 2004a. DOI: 10.1007/s00778-004-0132-6 7, 24

B. Babcock, M. Datar, and R. Motwani. Load shedding for aggregation queries over data streams. In *Proc. 20th Int. Conf. on Data Engineering*, pages 350–361, 2004b. 7, 35

S. Babu and J. Widom. StreaMon: an adaptive engine for stream query processing. In *Proc. ACM SIGMOD Int. Conf. on Management of Data*, pages 931–932, 2004. DOI: 10.1145/1007568.1007702 7, 36

S. Babu, R. Motwani, K. Munagala, I. Nishizawa, and J. Widom. Adaptive ordering of pipelined stream filters. In *Proc. ACM SIGMOD Int. Conf. on Management of Data*, pages 407–418, 2004a. DOI: 10.1145/1007568.1007615 30

S. Babu, U. Srivastava, and J. Widom. Exploiting k-constraints to reduce memory overhead in continuous queries over data streams. *ACM Trans. Database Syst.*, 29(3):545–580, Sep 2004b. 30

S. Babu, K. Munagala, J. Widom, and R. Motwani. Adaptive caching for continuous queries. In *Proc. 21st Int. Conf. on Data Engineering*, pages 118–129, 2005. DOI: 10.1109/ICDE.2005.15 36

Y. Bai, H. Thakkar, H. Wang, C. Luo, and C. Zaniolo. A data stream language and system designed for power and extensibility. In *Proc. 15th ACM Int. Conf. on Information and Knowledge Management*, pages 337–346, 2006. DOI: /10.1145/1183614.1183664 7, 21

H. Balakrishnan, M. Balazinska, D. Carney, U. Cetintemel, M. Cherniack, C. Convey, E. Galvez, J. Salz, M. Stonebraker, N. Tatbul, R. Tibbetts, and S. Zdonik. Retrospective on aurora. *VLDB J.*, 13(4):370–383, 2004. DOI: 10.1007/s00778-004-0133-5 6

M. Balazinska, H. Balakrishnan, S. Madden, and M. Stonebraker. Fault-tolerance in the borealis distributed stream processing system. In *Proc. ACM SIGMOD Int. Conf. on Management of Data*, pages 13–24, 2005. DOI: 10.1145/1066157.1066160 6

M. Balazinska, Y. Kwon, N. Kuchta, and D. Lee. Moirae: History-enhanced monitoring. In *Proc. 3rd Biennial Conf. on Innovative Data Systems Research*, pages 375–386, 2007. 14

M. Bateni, L. Golab, M. Hajiaghayi, and H. Karloff. Scheduling to minimize staleness and stretch in real-time data warehouses. In *Proc. 21st ACM Symp. on Parallel Algorithms and Architectures*, pages 29–38, 2009. DOI: 10.1145/1583991.1583998 44

P. Bizarro, S. Babu, D. DeWitt, and J. Widom. Content-based routing: Different plans for different data. In *Proc. 31st Int. Conf. on Very Large Data Bases*, pages 757–768, 2005. 36

A. Bulut and A. K. Singh. SWAT: Hierarchical stream summarization. In *Proc. 19th Int. Conf. on Data Engineering*, pages 303–314, 2003. 31

A. Bulut and A. K. Singh. A unified framework for monitoring data streams in real time. In *Proc. 21st Int. Conf. on Data Engineering*, pages 44–55, 2005. DOI: 10.1109/ICDE.2005.13 31

M. Cammert, J. Kramer, B. Seeger, and S. Vaupel. A cost-based approach to adaptive resource management in data stream systems. *IEEE Trans. Knowl. and Data Eng.*, 20(2):230–245, 2008. DOI: 10.1109/TKDE.2007.190686 29

D. Carney, U. Cetintemel, A. Rasin, S. Zdonik, M. Cherniack, and M. Stonebraker. Operator scheduling in a data stream manager. In *Proc. 29th Int. Conf. on Very Large Data Bases*, pages 838–849, 2003. 6, 24

S. Chakravarthy and Q. Jiang, editors. *Stream Data Processing: A Quality of Service Perspective.* Springer, 2009. 6

S. Chandrasekaran and M. J. Franklin. PSoup: a system for streaming queries over streaming data. *VLDB J.*, 12(2):140–156, Aug 2003. DOI: 10.1007/s00778-003-0096-y 31, 34, 36

S. Chandrasekaran and M. J. Franklin. Remembrance of streams past: overload-sensitive management of archived streams. In *Proc. 30th Int. Conf. on Very Large Data Bases*, pages 348–359, 2004. 7

S. Chandrasekaran, O. Cooper, A. Deshpande, M. J. Franklin, J. M. Hellerstein, W. Hong, S. Krishnamurthy, S. Madden, V. Raman, F. Reiss, and M. Shah. TelegraphCQ: Continuous dataflow processing for an uncertain world. In *Proc. 1st Biennial Conf. on Innovative Data Systems Research*, pages 269–280, 2003. 7

N. Chaudhry, K. Shaw, and M. Abdelguerfi, editors. *Stream Data Management*. Springer, 2005. DOI: 10.1007/b106968 6

J. Chen, D. DeWitt, F. Tian, and Y. Wang. NiagaraCQ: A scalable continuous query system for internet databases. In *Proc. ACM SIGMOD Int. Conf. on Management of Data*, pages 379–390, 2000. DOI: 10.1145/335191.335432 7, 33

J. Chen, D. DeWitt, and J. Naughton. Design and evaluation of alternative selection placement strategies in optimizing continuous queries. In *Proc. 18th Int. Conf. on Data Engineering*, pages 345–357, 2002. DOI: 10.1109/ICDE.2002.994749 7, 33

R. Cheng, B. Kao, S. Prabhakar, A. Kwan, and Y. Tu. Adaptive stream filters for entity-based queries with non-value tolerance. In *Proc. 31st Int. Conf. on Very Large Data Bases*, pages 37–48, 2005. 37

M. Cherniack and S. Zdonik. Stream-oriented query languages and operators. In L. Liu and M. T. Özsu, editors, *Encyclopedia of Database Systems*, pages 2848–2854. Springer, 2009. DOI: 10.1007/978-0-387-39940-9_368 22

J. Cho and H. Garcia-Molina. Synchronizing a database to im-prove freshness. In *Proc. ACM SIGMOD Int. Conf. on Management of Data*, pages 117–128, 2000. DOI: 10.1145/335191.335391 44

E. Cohen and M. Strauss. Maintaining time-decaying stream aggregates. In *Proc. ACM SIGACT-SIGMOD Symp. on Principles of Database Systems*, pages 223–233, 2003. DOI: 10.1016/j.jalgor.2005.01.006 33

S. Cohen. User-defined aggregate functions: bridging theory and practice. In *Proc. ACM SIGMOD Int. Conf. on Management of Data*, pages 49–60, 2006. DOI: 10.1145/1142473.1142480 31

G. Cormode and M. Garofalakis. Sketching streams through the net: Distributed approximate query tracking. In *Proc. 31st Int. Conf. on Very Large Data Bases*, pages 13–24, 2005. 37

G. Cormode and S. Muthukrishnan. An improved data stream summary: The count-min sketch and its applications. In *Proc. 6th Latin American Theor. Informatics Symp.*, pages 29–38, 2004. DOI: 10.1016/j.jalgor.2003.12.001 16

G. Cormode, T. Johnson, F. Korn, S. Muthukrishnan, I. Spatscheck, and D. Srivastava. Holistic UDAFs at streaming speeds. In *Proc. ACM SIGMOD Int. Conf. on Management of Data*, pages 35–46, 2004. DOI: 10.1145/1007568.1007575 7, 17

G. Cormode, M. Garofalakis, S. Muthukrishnan, and R. Rastogi. Holistic aggregates in a networked world: distributed tracking of approximate quantiles. In *Proc. ACM SIGMOD Int. Conf. on Management of Data*, pages 25–36, 2005. DOI: 10.1145/1066157.1066161 37

G. Cormode, S. Muthukrishnan, and W. Zhuang. What's different: distributed, continuous monitoring of duplicate-resilient aggregates on data streams. In *Proc. 22nd Int. Conf. on Data Engineering*, page 57, 2006. DOI: 10.1109/ICDE.2006.173 37

C. Cranor, T. Johnson, O. Spatscheck, and V. Shkapenyuk. Gigascope: A stream database for network applications. In *Proc. ACM SIGMOD Int. Conf. on Management of Data*, pages 647–651, 2003. DOI: 10.1145/872757.872838 7, 12, 19

Y. Cui and J. Widom. Practical lineage tracing in data warehouses. In *Proc. 16th Int. Conf. on Data Engineering*, pages 367–378, 2000. DOI: 10.1109/ICDE.2000.839437 41

A. Das, S. Ganguly, M. Garofalakis, and R. Rastogi. Distributed set-expression cardinality estimation. In *Proc. 30th Int. Conf. on Very Large Data Bases*, pages 312–323, 2004. 37

A. Das, J. Gehrke, and M. Riedewald. Semantic approximation of data stream joins. *IEEE Trans. Knowl. and Data Eng.*, 17(1):44–59, 2005. DOI: 10.1109/TKDE.2005.17 35

M. Datar, A. Gionis, P. Indyk, and R. Motwani. Maintaining stream statistics over sliding windows. In *Proc. 13th Annual ACM-SIAM Symp. on Discrete Algorithms*, pages 635–644, 2002. DOI: 10.1137/S0097539701398363 33

E. Demaine, A. Lopez-Ortiz, and J. I. Munro. Frequency estimation of internet packet streams with limited space. In *Proc. 10th Annual European Symp. on Algorithms*, pages 348–360, 2002. DOI: 10.1007/3-540-45749-6_33 16

A. Demers, J. Gehrke, M. Hong, M. Riedewald, and W. White. Towards expressive publish/subscribe systems. In *Advances in Database Technology, Proc. 10th Int. Conf. on Extending Database Technology*, pages 627–644, 2006. 34

A. Demers, J. Gehrke, B. Panda, M. Riedewald, V. Sharma, and W. White. Cayuga: a general purpose event monitoring system. In *Proc. 3rd Biennial Conf. on Innovative Data Systems Research*, pages 412–422, 2007. DOI: 10.1145/1247480.1247620 6

M. Denny and M. Franklin. Predicate result range caching for continuous queries. In *Proc. ACM SIGMOD Int. Conf. on Management of Data*, pages 646–657, 2005. DOI: 10.1145/1066157.1066231 33

M. Denny and M. Franklin. Operators for expensive functions in continuous queries. In *Proc. 22nd Int. Conf. on Data Engineering*, page 147, 2006. DOI: 10.1109/ICDE.2006.110 34

A. Deshpande. An initial study of overheads of Eddies. *ACM SIGMOD Rec.*, 33(1):44–49, 2004. DOI: 10.1145/974121.974129 36

A. Deshpande and J. Hellerstein. Lifting the burden of history from adaptive query processing. In *Proc. 30th Int. Conf. on Very Large Data Bases*, pages 948–959, 2004. 7, 36

L. Ding and E. Rundensteiner. Evaluating window joins over punctuated streams. In *Proc. 14th ACM Int. Conf. on Information and Knowledge Management*, pages 98–107, 2004. DOI: 10.1145/1031171.1031189 25

L. Ding, N. Mehta, E. Rundensteiner, and G. Heineman. Joining punctuated streams. In *Advances in Database Technology, Proc. 9th Int. Conf. on Extending Database Technology*, pages 587–604, 2004. 25

A. Dobra, M. Garofalakis, J. Gehrke, and R. Rastogi. Processing complex aggregate queries over data streams. In *Proc. ACM SIGMOD Int. Conf. on Management of Data*, pages 61–72, 2002. DOI: 10.1145/564691.564699 33

C. Estan and G. Varghese. New directions in traffic measurement and accounting. In *Proc. ACM Int. Conf. on Data Communication*, pages 323–336, 2002. DOI: 10.1145/964725.633056 16

F. Fabret, H.-A. Jacobsen, F. Llirbat, J. Pereira, K. Ross, and D. Shasha. Filtering algorithms and implementation for very fast publish/subscribe systems. In *Proc. ACM SIGMOD Int. Conf. on Management of Data*, pages 115–126, 2001. DOI: 10.1145/376284.375677 34

R. Fernandez-Moctezuma, K. Tufte, and J. Li. Inter-operator feedback in data stream management systems via punctuation. In *Proc. 4th Biennial Conf. on Innovative Data Systems Research*, 2009. 25

P. Flajolet and G. N. Martin. Probabilistic counting. In *Proc. IEEE Conf. on Foundations of Computer Science*, pages 76–82, 1983. 16

N. Folkert, A. Gupta, A. Witkowski, S. Subramanian, S. Bellamkonda, S. Shankar, T. Bozkaya, and L. Sheng. Optimizing refresh of a set of materialized views. In *Proc. 31st Int. Conf. on Very Large Data Bases*, pages 1043–1054, 2005. 40

M. Garofalakis, J. Gehrke, and R. Rastogi, editors. *Data Stream Management: Processing High-Speed Data Streams.* Springer, 2010. 6

B. Gedik, K.-L. Wu, P. S. Yu, and L. Liu. Adaptive load shedding for windowed stream joins. In *Proc. 14th ACM Int. Conf. on Information and Knowledge Management*, pages 171–178, 2005. DOI: 10.1145/1099554.1099587 35

B. Gedik, K.-L. Wu, and P. S. Yu. Efficient construction of compact shedding filters for data stream processing. In *Proc. 24th Int. Conf. on Data Engineering*, pages 396–405, 2008. DOI: 10.1109/ICDE.2008.4497448 34

T. Ghanem, W. Aref, and A. Elmagarmid. Exploiting predicate-window semantics over data streams. *ACM SIGMOD Rec.*, 35(1):3–8, 2006. DOI: 10.1145/1121995.1121996 11

T. Ghanem, M. Hammad, M. Mokbel, W. Aref, and A. Elmagarmid. Incremental evaluation of sliding-window queries over data streams. *IEEE Trans. Knowl. and Data Eng.*, 19(1):57–52, 2007. DOI: 10.1109/TKDE.2007.250585 7, 17, 26

T. Ghanem, A. Elmagarmid, P. Larson, and W. Aref. Supporting views in data stream management systems. *ACM Trans. Database Syst.*, 35(1), 2010. DOI: 10.1145/1670243.1670244 7, 12, 17, 20, 21

P. Gibbons. Distinct sampling for highly-accurate answers to distinct values queries and event reports. In *Proc. 27th Int. Conf. on Very Large Data Bases*, pages 541–550, 2001. 16

P. Gibbons and Y. Matias. New sampling-based summary statistics for improving approximate query answers. In *Proc. ACM SIGMOD Int. Conf. on Management of Data*, pages 331–342, 1998. DOI: 10.1145/276305.276334 16

P. Gibbons and S. Tirthapura. Distributed streams algorithms for sliding windows. In *Proc. 14th ACM Symp. on Parallel Algorithms and Architectures*, pages 63–72, 2002. DOI: 10.1145/564870.564880 33

A. C. Gilbert, Y. Kotidis, S. Muthukrishnan, and M. J. Strauss. Surfing wavelets on streams: One-pass summaries for approximate aggregate queries. In *Proc. 27th Int. Conf. on Very Large Data Bases*, pages 79–88, 2001. 10

A. C. Gilbert, Y. Kotidis, S. Muthukrishnan, and M. J. Strauss. How to summarize the universe: Dynamic maintenance of quantiles. In *Proc. 28th Int. Conf. on Very Large Data Bases*, pages 454–465, 2002. DOI: 10.1016/B978-155860869-6/50047-0 16

L. Golab. *Sliding window processing over data streams.* PhD thesis, David R. Cheriton School of Computer Science, University of Waterloo, Waterloo, Ontario, Canada, 2006. 17

L. Golab and M. T. Özsu. Issues in data stream management. *ACM SIGMOD Rec.*, 32(2):5–14, 2003a. DOI: 10.1145/776985.776986 6

L. Golab and M. T. Özsu. Processing sliding window multi-joins in continuous queries over data streams. In *Proc. 29th Int. Conf. on Very Large Data Bases*, pages 500–511, 2003b. 14, 30, 31

L. Golab and M. T. Özsu. Update-pattern aware modeling and processing of continuous queries. In *Proc. ACM SIGMOD Int. Conf. on Management of Data*, pages 658–669, 2005. DOI: 10.1145/1066157.1066232 17, 19, 26, 28

L. Golab, S. Garg, and M. T. Özsu. On indexing sliding windows over on-line data streams. In *Advances in Database Technology, Proc. 9th Int. Conf. on Extending Database Technology*, pages 712–729, 2004. 33

L. Golab, K. G. Bijay, and M. T. Özsu. On concurrency control in sliding window queries over data streams. In *Advances in Database Technology, Proc. 10th Int. Conf. on Extending Database Technology*, pages 608–626, 2006a. DOI: 10.1007/11687238_37 46

L. Golab, K. G. Bijay, and M. T. Özsu. Multi-query optimization of sliding window aggregates by schedule synchronization. In *Proc. 15th ACM Int. Conf. on Information and Knowledge Management*, pages 844–845, 2006b. DOI: 10.1145/1183614.1183759 33

L. Golab, P. Prahladka, and M. T. Özsu. Indexing time-evolving data with variable lifetimes. In *Proc. 18th Int. Conf. on Scientific and Statistical Database Management*, pages 265–274, 2006c. DOI: 10.1109/SSDBM.2006.29 42, 43

L. Golab, T. Johnson, N. Koudas, D. Srivastava, and D. Toman. Optimizing away joins on data streams. In *Proc. SSPS Workshop*, pages 48–57, 2008a. DOI: 10.1145/1379272.1379282 30

L. Golab, T. Johnson, and O. Spatscheck. Prefilter: predicate pushdown at streaming speeds. In *Proc. SSPS Workshop*, pages 29–37, 2008b. DOI: 10.1145/1379272.1379280 7, 30, 34

L. Golab, T. Johnson, J. S. Seidel, and V. Shkapenyuk. Stream warehousing with datadepot. In *Proc. ACM SIGMOD Int. Conf. on Management of Data*, pages 847–854, 2009a. DOI: 10.1145/1559845.1559934 2, 7, 40, 41

L. Golab, T. Johnson, and V. Shkapenyuk. Scheduling updates in a real-time stream warehouse. In *Proc. 25th Int. Conf. on Data Engineering*, pages 1207–1210, 2009b. DOI: 10.1109/ICDE.2009.202 7, 44, 45, 46

J. Greenwald and F. Khanna. Space efficient on-line computation of quantile summaries. In *Proc. ACM SIGMOD Int. Conf. on Management of Data*, pages 58–66, 2001. DOI: 10.1145/376284.375670 16

X. Gu, P. Yu, and H. Wang. Adaptive load diffusion for multiway windowed stream joins. In *Proc. 23rd Int. Conf. on Data Engineering*, pages 146–155, 2007. DOI: 10.1109/ICDE.2007.367860 36

M. Hammad, M. J. Franklin, W. Aref, and A. Elmagarmid. Scheduling for shared window joins over data streams. In *Proc. 29th Int. Conf. on Very Large Data Bases*, pages 297–308, 2003. 33

M. Hammad, M. Mokbel, M. Ali, W. Aref, A. Catlin, A. Elmagarmid, M. Eltabakh, M. Elfeky, T. Ghanem, R. Gwadera, I. Ilyas, M. Marzouk, and X. Xiong. Nile: a query processing engine for data streams. In *Proc. 20th Int. Conf. on Data Engineering*, page 851, 2004. DOI: 10.1109/ICDE.2004.1320080 7

E. Hanson, C. Carnes, L. Huang, M. Konyala, and L. Noronha. Scalable trigger processing. In *Proc. 15th Int. Conf. on Data Engineering*, pages 266–275, 1999. 34

J. M. Hellerstein, P. Haas, and H. Wang. Online aggregation. In *Proc. ACM SIGMOD Int. Conf. on Management of Data*, pages 171–182, 1997. DOI: 10.1145/253262.253291 13

M. Hoffmann, S. Muthukrishnan, and R. Raman. Streaming algorithms for data in motion. In *Proc. Int. Symp. on Combinatorics, Algorithms, Probabilistic and Experimental Methodologies*, pages 294–304, 2007. DOI: 10.1007/978-3-540-74450-4_27 10

M. Hong, M. Riedewald, C. Koch, J. Gehrke, and A. Demers. Rule-based multi-query optimization. In *Advances in Database Technology, Proc. 12th Int. Conf. on Extending Database Technology*, pages 120–131, 2009. DOI: 10.1145/1516360.1516376 33

J.-H. Hwang, Y. Xing, U. Cetintemel, and S. Zdonik. A cooperative, self-configuring high-availability solution for stream processing. In *Proc. 23rd Int. Conf. on Data Engineering*, pages 176–185, 2007. DOI: 10.1109/ICDE.2007.367863 6

A. Jain, E. Y. Chang, and Y-F. Wang. Adaptive stream resource management using Kalman Filters. In *Proc. ACM SIGMOD Int. Conf. on Management of Data*, pages 11–22, 2004. DOI: 10.1145/1007568.1007573 37

N. Jain, S. Mishra, A. Srinivasan, J. Gehrke, J. Widom, H. Balakrishnan, U. Cetintemel, M. Cherniack, R. Tibbetts, and S. Zdonik. Towards a streaming SQL standard. In *Proc. 34th Int. Conf. on Very Large Data Bases*, pages 1379–1390, 2008. DOI: 10.1145/1454159.1454179 10

T. Johnson, S. Muthukrishnan, and I. Rozenbaum. Sampling algorithms in a stream operator. In *Proc. ACM SIGMOD Int. Conf. on Management of Data*, pages 1–12, 2005a. DOI: 10.1145/1066157.1066159 7, 16, 22

T. Johnson, S. Muthukrishnan, V. Shkapenyuk, and O. Spatscheck. A heartbeat mechanism and its application in Gigascope. In *Proc. 31st Int. Conf. on Very Large Data Bases*, pages 1079–1088, 2005b. 7, 25

T. Johnson, S. Muthukrishnan, O. Spatscheck, and D. Srivastava. Streams, security and scalability. In *Proc. 19th Annual IFIP Conf. on Data and Applications Security, LNCS 3654*, pages 1–15, 2005c. DOI: 10.1007/11535706_1 7, 35

T. Johnson, S. Muthukrishnan, V. Shkapenyuk, and O. Spatscheck. Query-aware partitioning for monitoring massive network data streams. In *Proc. ACM SIGMOD Int. Conf. on Management of Data*, pages 1135–1146, 2008. DOI: 10.1145/1376616.1376730 1, 7, 36

J. Kang, J. Naughton, and S. Viglas. Evaluating window joins over unbounded streams. In *Proc. 19th Int. Conf. on Data Engineering*, pages 341–352, 2003. DOI: 10.1145/1031171.1031189 14, 29

R. Keralapura, G. Cormode, and J. Ramamirtham. Communication-efficient distributed monitoring of threshold counts. In *Proc. ACM SIGMOD Int. Conf. on Management of Data*, pages 289–300, 2006. DOI: 10.1145/1142473.1142507 37

J. Krämer. *Continuous queries over data streams - Semantics and implementation*. PhD thesis, University of Marburg, Marburg, Germany, 2007. 7

J. Krämer and B. Seeger. PIPES - a public infrastructure for processing and exploring streams. In *Proc. ACM SIGMOD Int. Conf. on Management of Data*, pages 925–926, 2004. DOI: 10.1145/1007568.1007699 7

J. Krämer and B. Seeger. Semantics and implementation of continuous sliding window queries over data streams. *ACM Trans. Database Syst.*, 34(1), 2009. DOI: 10.1145/1508857.1508861 7, 12, 29

S. Krishnamurthy, M. Franklin, J. Hellerstein, and G. Jacobson. The case for precision sharing. In *Proc. 30th Int. Conf. on Very Large Data Bases*, pages 972–986, 2004. 34

S. Krishnamurthy, C. Wu, and M. Franklin. On-the-fly sharing for streamed aggregation. In *Proc. ACM SIGMOD Int. Conf. on Management of Data*, pages 623–634, 2006. DOI: 10.1145/1142473.1142543 33, 34, 41

A. Labrinidis and N. Roussopoulos. Update propagation strat-egies for improving the quality of data on the web. In *Proc. 27th Int. Conf. on Very Large Data Bases*, pages 391–400, 2001. 44

Y-N. Law, H. Wang, and C. Zaniolo. Query languages and data models for database sequences and data streams. In *Proc. 30th Int. Conf. on Very Large Data Bases*, pages 492–503, 2004. 13, 17

L. K. Lee and H. F. Ting. A simpler and more efficient deterministic scheme for finding frequent items over sliding windows. In *Proc. ACM SIGACT-SIGMOD Symp. on Principles of Database Systems*, pages 290–297, 2006. DOI: 10.1145/1142351.1142393 33

F. Li, C. Chang, G. Kollios, and A. Bestavros. Characterizing and exploiting reference locality in data stream applications. In *Proc. 22nd Int. Conf. on Data Engineering*, page 81, 2006a. DOI: 10.1109/ICDE.2006.33 35

H-G. Li, S. Chen, J. Tatemura, D. Agrawal, K. S. Candan, and W-P. Hsiung. Safety guarantee of continuous join queries over punctuated data streams. In *Proc. 32nd Int. Conf. on Very Large Data Bases*, 2006b. 25

J. Li, D. Maier, K. Tufte, V. Papadimos, and P. Tucker. Semantics and evaluation techniques for window aggregates in data streams. In *Proc. ACM SIGMOD Int. Conf. on Management of Data*, pages 311–322, 2005a. DOI: 10.1145/1066157.1066193 25

J. Li, D. Maier, K. Tufte, V. Papadimos, and P. Tucker. No pane, no gain: efficient evaluation of sliding-window aggregates over data streams. *ACM SIGMOD Rec.*, 34(1):39–44, 2005b. DOI: 10.1145/1058150.1058158 33

J. Li, K. Tufte, V. Shkapenyuk, V. Papadimos, T. Johnson, and D. Maier. Out-of-order processing: a new architecture for high-performance stream systems. In *Proc. 34th Int. Conf. on Very Large Data Bases*, pages 274–288, 2008. DOI: 10.1145/1453856.1453890 25

H.-S. Lim, J.-G. Lee, M.-J. Lee, K.-Y. Whang, and I.-Y. Song. Continuous query processing in data streams using duality of data and queries. In *Proc. ACM SIGMOD Int. Conf. on Management of Data*, pages 313–324, 2006. DOI: 10.1145/1142473.1142509 34

X. Lin, H. Lu, J. Xu, and J. Xu Yu. Continuously maintaining quantile summaries of the most recent N elements over a data stream. In *Proc. 20th Int. Conf. on Data Engineering*, pages 362–373, 2004. DOI: 10.1109/ICDE.2004.1320011 33

X. Lin, Y. Yuan, W. Wang, and H. Lu. Stabbing the sky: Efficient skyline computation over sliding windows. In *Proc. 21st Int. Conf. on Data Engineering*, pages 502–513, 2005. DOI: 10.1109/ICDE.2005.137 28

B. Liu, Y. Zhu, M. Jbantova, B. Momberger, and E. Rundensteiner. A dynamically adaptive distributed system for processing complex continuous queries. In *Proc. 31st Int. Conf. on Very Large Data Bases*, pages 1338–1341, 2005. 7, 36

B. Liu, Y. Zhu, and E. Rundensteiner. Run-time operator state spilling for memory intensive long running queries. In *Proc. ACM SIGMOD Int. Conf. on Management of Data*, pages 347–358, 2006a. DOI: 10.1145/1142473.1142513 35

H. Liu, Y. Lu, J. Han, and J. He. Error-adaptive and time-aware maintenance of frequency counts over data streams. In *Proc. 7th Int. Conf. on Web-Age Information Management*, pages 484–495, 2006b. DOI: 10.1007/11775300_41 16

L. Ma, S. Viglas, M. Li, and Q. Li. Stream operators for querying data streams. In *Proc. 6th Int. Conf. on Web-Age Information Management*, pages 404–415, 2005. DOI: 10.1007/11563952_36 12

S. Madden, M. J. Franklin, J. M. Hellerstein, and W. Hong. TAG: a tiny aggregation service for ad-hoc sensor networks. In *Proc. 5th USENIX Symp. on Operating System Design and Implementation*, 2002a. DOI: 10.1145/844128.844142 37

S. Madden, M. Shah, J. Hellerstein, and V. Raman. Continuously adaptive continuous queries over streams. In *Proc. ACM SIGMOD Int. Conf. on Management of Data*, pages 49–60, 2002b. DOI: 10.1145/564691.564698 34, 36, 41

A. Manjhi, V. Shkapenyuk, K. Dhamdhere, and C. Olston. Finding (recently) frequent items in distributed data streams. In *Proc. 21st Int. Conf. on Data Engineering*, pages 767–778, 2005. DOI: 10.1109/ICDE.2005.68 37

G. Manku, S. Rajagopalan, and B. Lindsay. Random sampling techniques for space efficient online computation of order statistics of large datasets. In *Proc. ACM SIGMOD Int. Conf. on Management of Data*, pages 251–262, 1999. DOI: 10.1145/304181.304204 16

G. S. Manku and R. Motwani. Approximate frequency counts over data streams. In *Proc. 28th Int. Conf. on Very Large Data Bases*, pages 346–357, 2002. DOI: 10.1016/B978-155860869-6/50038-X 16

A. Metwally, D. Agrawal, and A. El Abbadi. Efficient computation of frequent and top-k elements in data streams. In *Proc. 10th Int. Conf. on Database Theory*, pages 398–412, 2005. DOI: 10.1007/b104421 16

R. Motwani, J. Widom, A. Arasu, B. Babcock, S. Babu, M. Datar, G. Manku, C. Olston, J. Rosenstein, and R. Varma. Query processing, approximation, and resource management in a data stream management system. In *Proc. 1st Biennial Conf. on Innovative Data Systems Research*, pages 245–256, 2003. 7, 16

K. Mouratidis, S. Bakiras, and D. Papadias. Continuous monitoring of top-k queries over sliding windows. In *Proc. ACM SIGMOD Int. Conf. on Management of Data*, pages 635–646, 2006. DOI: 10.1145/1142473.1142544 28

S. Muthukrishnan. Data streams: Algorithms and applications. *Foundations and Trends in Theoretical Computer Science*, 1(2), 2005. 6, 10, 17

S. Nath, P. Gibbons, S. Seshan, and Z. Anderson. Synopsis diffusion for robust aggregation in sensor networks. In *Proc. 2nd Int. Conf. on Embedded Networked Sensor Systems*, pages 250–262, 2004. DOI: 10.1145/1031495.1031525 37

R. Nehme, E. Rundensteiner, and E. Bertino. A security punctuation framework for enforcing access control on streaming data. In *Proc. 24th Int. Conf. on Data Engineering*, pages 406–415, 2008. DOI: 10.1109/ICDE.2008.4497449 25

R. Nehme, E. Rundensteiner, and E. Bertino. Tagging stream data for rich real-time services. In *Proc. 35th Int. Conf. on Very Large Data Bases*, pages 73–84, 2009. 25

C. Olston, J. Jiang, and J. Widom. Adaptive filters for continuous queries over distributed data streams. In *Proc. ACM SIGMOD Int. Conf. on Management of Data*, pages 563–574, 2003. DOI: 10.1145/872757.872825 37

A. Pavan and S. Tirthapura. Range-efficient computation of F_0 over massive data streams. In *Proc. 21st Int. Conf. on Data Engineering*, pages 32–43, 2005. DOI: 10.1109/ICDE.2005.118 16

N. Polyzotis, S. Skiadopoulos, P. Vassiliadis, A. Simitsis, and N. Frantzell. Meshing streaming updates with persistent data in an active data warehouse. *IEEE Trans. Knowl. and Data Eng.*, 20 (7):976–991, 2008. DOI: 10.1109/TKDE.2008.27 2, 39

L. Qiao, D. Agrawal, and A. El Abbadi. Supporting sliding window queries for continuous data streams. In *Proc. 15th Int. Conf. on Scientific and Statistical Database Management*, pages 85–94, 2003. DOI: 10.1109/SSDM.2003.1214970 33

V. Raman, A. Deshpande, and J. Hellerstein. Using state modules for adaptive query processing. In *Proc. 19th Int. Conf. on Data Engineering*, pages 353–364, 2003. DOI: 10.1109/ICDE.2003.1260805 36

F. Reiss and J. Hellerstein. Data triage: an adaptive architecture for load shedding in telegraphCQ. In *Proc. 21st Int. Conf. on Data Engineering*, pages 155–156, 2005. DOI: 10.1109/ICDE.2005.44 7, 35

E. Rundensteiner, L. Ding, T. Sutherland, Y. Zhu, B. Pielech, and N. Mehta. CAPE: continuous query engine with heterogeneous-grained adaptivity. In *Proc. 30th Int. Conf. on Very Large Data Bases*, pages 1353–1356, 2004. 7

E. Ryvkina, A. Maskey, I. Adams, B. Sandler, C. Fuchs, M. Cherniack, and S. Zdonik. Revision processing in a stream processing engine: A high-level design. In *Proc. 22nd Int. Conf. on Data Engineering*, page 141, 2006. DOI: 10.1109/ICDE.2006.130 10

A. Salehi, M. Riahi, S. Michel, and K. Aberer. Knowing when to slide - efficient scheduling for sliding window processing. In *Proc. 10th Int. Conf. on Mobile Data Management*, pages 202–211, 2009. DOI: 10.1109/MDM.2009.31 33

M. A. Shah, J. M. Hellerstein, S. Chandrasekaran, and M. J. Franklin. Flux: An adaptive partitioning operator for continuous query systems. In *Proc. 19th Int. Conf. on Data Engineering*, pages 25–36, 2003. DOI: 10.1109/ICDE.2003.1260779 36

M. Sharaf, P. Chrysanthis, A. Labrinidis, and K. Pruhs. Algorithms and metrics for processing multiple heterogeneous continuous queries. *ACM Trans. Database Syst.*, 33(1), 2008. DOI: 10.1145/1331904.1331909 24

I. Sharfman, A. Schuster, and D. Keren. A geometric approach to monitoring threshold functions over distributed data streams. *ACM Trans. Database Syst.*, 4(32), 2007. DOI: 10.1145/1292609.1292613 37

N. Shivakumar and H. García-Molina. Wave-indices: indexing evolving databases. In *Proc. ACM SIGMOD Int. Conf. on Management of Data*, pages 381–392, 1997. DOI: 10.1145/253260.253349 40

U. Srivastava and J. Widom. Flexible time management in data stream systems. In *Proc. ACM SIGACT-SIGMOD Symp. on Principles of Database Systems*, pages 263–274, 2004a. DOI: 10.1145/1055558.1055596 25

U. Srivastava and J. Widom. Memory-limited execution of windowed stream joins. In *Proc. 30th Int. Conf. on Very Large Data Bases*, pages 324–335, 2004b. 35

M. Stonebraker, U. Cetintemel, and S. Zdonik. The 8 requirements of real-time stream processing. *ACM SIGMOD Rec.*, 34(4):42–47, 2005. DOI: 10.1145/1107499.1107504 6

Y. Tao and D. Papadias. Maintaining sliding window skylines on data streams. *IEEE Trans. Knowl. and Data Eng.*, 18(3):377–391, 2006. DOI: 10.1109/TKDE.2006.48 28

N. Tatbul and S. Zdonik. Window-aware load shedding for aggregation queries over data streams. In *Proc. 32nd Int. Conf. on Very Large Data Bases*, pages 799–810, 2006. 35

N. Tatbul, U. Cetintemel, S. Zdonik, M. Cherniack, and M. Stonebraker. Load shedding in a data stream manager. In *Proc. 29th Int. Conf. on Very Large Data Bases*, pages 309–320, 2003. 6, 34

N. Tatbul, U. Cetintemel, and S. Zdonik. Staying FIT: Efficient load shedding techniques for distributed stream processing. In *Proc. 33rd Int. Conf. on Very Large Data Bases*, pages 159–170, 2007. 7

F. Tian and D. DeWitt. Tuple routing strategies for distributed Eddies. In *Proc. 29th Int. Conf. on Very Large Data Bases*, pages 333–344, 2003. 36

W. H. Tok and S. Bressan. Efficient and adaptive processing of multiple continuous queries. In *Advances in Database Technology, Proc. 8th Int. Conf. on Extending Database Technology*, pages 215–232, 2002. DOI: 10.1007/3-540-45876-X_15 36

Y.-C. Tu, S. Liu, S. Prabhakar, and B. Yao. Load shedding in stream databases: a control-based approach. In *Proc. 32nd Int. Conf. on Very Large Data Bases*, pages 787–798, 2006. 35

P. Tucker, D. Maier, T. Sheard, and L. Faragas. Exploiting punctuation semantics in continuous data streams. *IEEE Trans. Knowl. and Data Eng.*, 15(3):555–568, 2003. DOI: 10.1109/TKDE.2003.1198390 25

S. Viglas and J. Naughton. Rate-based query optimization for streaming information sources. In *Proc. ACM SIGMOD Int. Conf. on Management of Data*, pages 37–48, 2002. DOI: 10.1145/564691.564697 24, 31

S. Viglas, J. Naughton, and J. Burger. Maximizing the output rate of multi-join queries over streaming information sources. In *Proc. 29th Int. Conf. on Very Large Data Bases*, pages 285–296, 2003. 14, 31

S. Wang and E. Rundensteiner. Scalable stream join processing with expensive predicates: workload distribution and adaptation by time-slicing. In *Advances in Database Technology, Proc. 12th Int. Conf. on Extending Database Technology*, pages 299–310, 2009. DOI: 10.1145/1516360.1516396 7

S. Wang, E. Rundensteiner, S. Ganguly, and S. Bhatnagar. State-slice: New paradigm of multi-query optimization of window-based stream queries. In *Proc. 32nd Int. Conf. on Very Large Data Bases*, pages 619–630, 2006. 34

J. Widom. Trio: A system for integrated management of data, accuracy, and lineage. In *Proc. 2nd Biennial Conf. on Innovative Data Systems Research*, pages 262–276, 2005. 41

A. Wilschut and P. Apers. Dataflow query execution in a parallel main-memory environment. In *Proc. 1st Int. Conf. on Parallel and Distributed Information Systems*, pages 68–77, 1991. DOI: 10.1109/PDIS.1991.183069 13

E. Wu, Y. Diao, and S. Rizvi. High-performance complex event processing over streams. In *Proc. ACM SIGMOD Int. Conf. on Management of Data*, pages 407–418, 2006. DOI: 10.1145/1142473.1142520 6

K.-L. Wu, S.-K. Chen, and P. Yu. Interval query indexing for efficient stream processing. In *Proc. 13th ACM Int. Conf. on Information and Knowledge Management*, pages 88–97, 2004. DOI: 10.1145/1031171.1031188 34

J. Xie, J. Yang, and Y. Chen. On joining and caching stochastic streams. In *Proc. ACM SIGMOD Int. Conf. on Management of Data*, pages 359–370, 2005. DOI: 10.1145/1066157.1066199 35

Y. Xing, S. Zdonik, and J.-H. Hwang. Dynamic load distribution in the borealis stream processor. In *Proc. 21st Int. Conf. on Data Engineering*, pages 791–802, 2005. DOI: 10.1109/ICDE.2005.53 7, 36

Y. Xing, J-H. Hwang, U. Cetintemel, and S. Zdonik. Providing resiliency to load variations in distributed stream processing. In *Proc. 32nd Int. Conf. on Very Large Data Bases*, pages 775–786, 2006. 7, 36

J. Xu, X. Lin, and X. Zhou. Space efficient quantile summary for constrained sliding windows on a data stream. In *Proc. 5th Int. Conf. on Web-Age Information Management:*, pages 34–44, 2004. 33

Y. Yang and D. Papadias. Just-in-time processing of continuous queries. In *Proc. 24th Int. Conf. on Data Engineering*, pages 1150–1159, 2008. DOI: 10.1109/ICDE.2008.4497524 31

R. Zhang, N. Koudas, B. C. Ooi, and D. Srivastava. Multiple aggregations over data streams. In *Proc. ACM SIGMOD Int. Conf. on Management of Data*, pages 299–310, 2005. DOI: 10.1145/1066157.1066192 33

X. Zhang and D. Shasha. Better burst detection. In *Proc. 22nd Int. Conf. on Data Engineering*, page 146, 2006. DOI: 10.1109/ICDE.2006.30 31

Y. Zhu and D. Shasha. StatStream: Statistical monitoring of thousands of data streams in real time. In *Proc. 28th Int. Conf. on Very Large Data Bases*, pages 358–369, 2002. DOI: 10.1016/B978-155860869-6/50039-1 33

Y. Zhu and D. Shasha. Efficient elastic burst detection in data streams. In *Proc. 9th ACM SIGKDD Int. Conf. on Knowledge Discovery and Data Mining*, pages 336–345, 2003. DOI: 10.1145/956750.956789 31

Y. Zhu, E. Rundensteiner, and G. Heineman. Dynamic plan migration for continuous queries over data streams. In *Proc. ACM SIGMOD Int. Conf. on Management of Data*, pages 431–442, 2004. DOI: 10.1145/1007568.1007617 7, 36

Authors' Biographies

LUKASZ GOLAB

Lukasz Golab is a Senior Member of Research Staff at AT&T Labs - Research. His research interests include data stream management, data warehousing, and data quality analysis. He received his PhD from the University of Waterloo in 2006, winning the Alumni Gold Medal as the top PhD student at the university.

M. TAMER ÖZSU

M. Tamer Özsu is a professor of computer science and University Research Chair at the University of Waterloo. Between January 2007 and July 2010, he was the Director of the David R. Cheriton School of Computer Science. Prior to his current position, he was with the Department of Computing Science of the University of Alberta between 1984 and 2000. He holds a PhD from Ohio State University.

Dr. Özsu's current research covers distributed data management, focusing on data stream systems, distributed XML processing, and data integration, and multimedia data management.

He is a Fellow of ACM, a Senior Member of IEEE, and a member of Sigma Xi; he is the recipient of 2006 ACM SIGMOD Contributions Award and the 2008 Ohio State University College of Engineering Distinguished Alumnus Award.